Sustainable Mountain Development

Jack D. Ives

Sustainable Mountain Development

Getting the facts right

Second Edition

 Springer

Jack D. Ives
Carleton University
Ottawa, ON, Canada

ISBN 978-3-030-96031-5   ISBN 978-3-030-96029-2   (eBook)
https://doi.org/10.1007/978-3-030-96029-2

This Springer imprint is published by the registered company Springer Nature Switzerland AG
The registered company address is: Gewerbestrasse 11, 6330 Cham, Switzerland

To

_____

Walther Manshard

Former Vice-Rector, United Nations University

for inspiring leadership

and steadfast friendship

Facts do not cease to exist because they are ignored.

- Aldous Huxley

Also by Jack D. Ives

The Land Beyond: A memoir. 2010, Alaska University Press, Fairbanks, U.S.A., 186 pp.

Skaftafell in Iceland: A Thousand Years of Change. 2007, Ormstunga, Reykjavik, Iceland, 256 pp.

Himalayan Perceptions: Environmental change and the well-being of mountain peoples. 2004, Routledge, London and New York, 284 pp. 2nd edition, 2006, Himalayan Assoc. Advancement Science, Kathmandu.

Mountains of the World: A Global Priority. 1997, co-editor and author, with Bruno Messerli. Parthenon Press, London and New York. (editions in Russian, French, Spanish, and Italian), 495 pp.

Polar Regions (The Illustrated Library of the World). 1995, Editor, with David Sugden. Readers' Digest, Australia Pty. Ltd., NSW., 160 pp.

Mountains (The Illustrated Library of the World). 1994, Editor, Rodale Press, Emmaus, USA, 160 pp.

The Himalayan Dilemma: Reconciling development and conservation. 1989, with Bruno Messerli. Routledge, London and New York, 295 pp.

Arctic and Alpine Environments. 1974, co-editor with Roger G. Barry. Methuen, London and New York, 999 pp.

Founding Editor of quarterly journals

Arctic, Antarctic, and Alpine Research, 1969

Mountain Research and Development, 1981

About the Author

Born in Grimsby, England, in 1931, the author graduated in geography from the University of Nottingham. He organized and led the university's first Arctic expeditions to Iceland (1952, 1953, 1954). In September, 1954, he married Pauline A. Cordingley, and they emigrated to Canada. Jack obtained his doctorate from McGill University, Montreal, and served as director of the McGill Sub-Arctic Research Laboratory where he initiated an extensive field research programme. From 1960 to 1967 he was assistant director and director of the federal government's Geographical Branch, Ottawa, and coordinated a series of seven interdisciplinary expeditions to Baffin Island. From 1967 to 1979 he served as director of the Institute of Arctic and Alpine Research, University of Colorado, Boulder, and professor of mountain geography from 1967 to 1989. During this period he founded and edited the quarterly journals, *Arctic and Alpine Research* (1969) and, with Pauline, *Mountain Research and Development* (1981). His final appointment was with the University of California, Davis (1989 to 1997). After retiring, he returned to Ottawa, Canada, where he was appointed Honorary Research Professor by Carleton University.

Jack was President of the Commission on Mountain Geoecology of the International Geographical Union for much of the period 1972 to 2000, Chair of the International Working Group of UNESCO's Man and the Biosphere (MAB) Project 6 – Mountains (1974–1977) and Coordinator of UNU's Project on Mountain Geoecology and Sustainable Development (1978–2000). He was awarded a John Simon Guggenheim Memorial Fellowship (1976–1977) tenable at Bern University, Switzerland, and was officially recognized by the Governor of the State of Colorado for his contributions to Rocky Mountain environmental problems. His other awards include: the King Albert 1st Gold Medal (2002); the Patron's Medal of the Royal Geographical Society (2006); and Iceland's Knight's Cross of the Order of the Falcon (2007).

He has published extensively on Arctic and Mountain topics and supervised over 50 Masters and doctoral candidates, and 12 post-docs. In December 2001 he gave a keynote address, representing the Rector of United Nations University, Professor Dr Hans J. A. van Ginkel, during the United Nations designation of 2002 as the International Year of Mountains. Jack and Pauline have a daughter, three sons, and five grandchildren.

Skaftafell and Iceland's highest mountain, Hvannadalshnukur, taken from summit of the Jökulfell in late evening. General setting of the University of Nottingham student expeditions, and the undergraduate experience that first impressed us how knowledgeable the local mountain farmers were about the nuances of their environment. (This photograph and Figs 1, 2 and 4 are included as they are intended to represent the author's earliest stages of involvement with mountains: Iceland-1952 to 1954; Labrador-Ungava-1955 to 1960; Baffin Island-1961 to 1967; and the Colorado Rockies-1968 to 1980).

Table of Contents

Preface

The first edition (2013) described the almost simultaneous origin of the UNESCO Man and the Biosphere Programme's Project on mountains (MAB-6) and the International Geographical Union's Commission on High Altitude Geoecology. The combination was the result of the efforts of a small group of academics devoted to mountains. This quickly led to an alliance with the newly created United Nations University, its project on "mountains", and the foundation of the International Mountain Society (1981) with its quarterly journal, *Mountain Research and Development*. Thus, there arose an indelible commitment to improving the well-being of poor mountain people together with their environment. Inevitably, this led to a challenge of a number of major "development" programmes. Several of these multi-million dollar projects short-circuited the mountain people, especially as they were often based on totally inadequate data ("mountain myths"). The initial major endeavour was to challenge what we referred to as "The Theory of Himalayan Environmental Degradation". The assumption that the Himalaya would rapidly become virtually deforested because of misuse by poor mountain farmers was totally overthrown after extensive field research (Chapter 10).

Our efforts expanded into the international political arena. The resulting expansion of support internationally, assisted by Maurice Strong, who became the Secretary General of the first "Earth Summit" (Rio de Janeiro, 1992), led to the UN's declaration of 2002 as the International Year of Mountains.

Following the 1992 Rio Earth Summit, UN FAO became the focal point for fostering mountain awareness. Since the publication of the first edition, it seems that the international approach to mountain environmental and people concerns has changed focus and has expanded worldwide. This fortunate development is powerfully illustrated in the new Epilogue for which I am indebted to my colleague, Martin Price.

The first edition was printed in Kathmandu and depended on the tireless work and support of another colleague, Kumar Mainali. He actually created the book and ensured an impressive array of more than a hundred colour photographs. Yet the actual distribution of the first edition was somewhat limited. Combined with the extensive and worldwide development of commitment to mountain people and their environment (see Epilogue by Martin Price), it was realized that this second edition was necessary, especially as many of the "mountain myths" are still perpetuated by elements of the news media.

For this new edition, I am greatly indebted to Aaron Schiller of Springer Environmental for the quality of production, achieved also by continued assistance from Kumar Mainali. We hope that its wider distribution will effectively challenge the post-2002, reduced yet continuing and seemingly undying news media attachment to "catastrophe" that has little factual support.

Jack D. Ives
1st November, 2021

Acknowledgements

The events covered by this book extend from 1964 virtually all the way to the present. It is inevitable, therefore, that my profound gratitude for many kinds of assistance and encouragement is extensive. The many individuals who played indispensable roles are mentioned as and when their participation occurred within the related chapters. Nevertheless, many of them must be further acknowledged in this overall retrospective of those who shared the mountain journey with me.

Carl Troll was the dominant figure during the earliest stages. It was his determination to establish the International Geographical Union (IGU) Commission on High-Altitude Geoecology that provided the intellectual spark. His friendship, inspiration, and confidence, and his very internationalization of what became "mountain geoecology" brought together a special group of academics embracing Europe, North and South America, Asia, and Africa. It also led to my first meeting with Bruno Messerli, to whom I owe an enormous debt. Bruno became my friend, confidant, and partner, and without him this long mountain adventure may never have occurred.

Walther Manshard, both as senior UNESCO officer and Vice-Rector of the United Nations University, and Jim Harrison, as Director of the Geological Survey of Canada and Assistant Director-General of UNESCO, pulled me into the UNU experiment in applied mountain geoecology without which none of the middle part of the mountain journey would have occurred. Frank Davidson, Francesco di Castri, Walter Moser, Gerardo Budowski, Klaus Lampe, Paul Baker, Corneille Jest, Mike Thompson, Larry Hamilton, Sun Honglie, David Griffin were all major figures. Their encouragement and intellectual challenge helped lay the foundation for the essential scepticism required for framing the vital questions needed to defeat the widely endorsed "mountain myths" and to create the sense of urgency to "get the facts right."

Maurice F. Strong played a dominant role from 1982 onward. Given his extensive responsibilities, especially his pivotal role in bringing the 1992 Rio Earth Summit to fruition, he still found time to provide seemingly constant personal attention and encouragement for our relatively small part in his remarkable world environmental whole.

Roger G. Barry has been an invaluable companion in the journey, from our first winter together at the McGill Sub-Arctic Research Laboratory in central Labrador-Ungava (1957–1958), via the Geographical Branch in Ottawa, to the Institute of Arctic and Alpine Research, Colorado, and, together with Mischa Plam, in the establishment of the International Mountain Society and its quarterly journal *Mountain Research and Development.* Roger has been totally supportive, loyal in very difficult circumstances, and a lifelong friend. He has set a hard-to-surpass example of professorial performance over a half-century.

Many colleagues, former graduate students, and post-doctoral fellows have provided the wherewithal for an effective and exacting series of mountain research expeditions: Don Alford, Yuri Badenkov, Khadga Basnet, Markus Bichsel, Barry Bishop, Inger Marie Bjønness, Barbara Brower, Muazama Burkhanova, Alton Byers, Elizabeth Byers, Nel Caine, Chen Baolin, Chen Chuanyou, Steve Cunha, D.N.S. Dhakal, Erwin Frei, Yang Fuquan, Wu Ga, Sumitra M. Gurung, Heinrich Hafner, Jörg Hanisch, Carol Harden, Thomas Hofer, Hans Hurni, Qiu Jiaqi, Kirsten Johnson, Du Juan, Narendra Khanal, Hans Kienholz, Liao Jungua, Liu Lanhui, Mu Liqin, Lee MacDonald, Janet D. Momson, Pradeep K. Mool, Elizabeth Ann Olson, Tjerk Peters, Guy Schneider, Kamal K. Shrestha, Seth Sicroff, Christoph Stadel, Ann Stettler, Margaret (Peggy) Swain, Lindsey Swope, Rabindra Tamrakar, Colin E. Thorn, Zuo Ting, Daniel Vuichard, Teiji Watanabe, Li Wenhua, Feng Zhao, Yang Zhouhuai, Yao Zhiyun, He Yaohua, Zhang Yongzu, and Markus Zimmermann.

Important organizational and intellectual contributions were made by Sunderlal Bahuguna, Jayanta Bandyopadhyay, Innokentiy Gerasimov, Jean-François Giovannini, Ruedi Hogger, Lise Grande, Fausto Maldonado, Nicholas Menzies, Fritz Müller, Huang Pingwei, Jane Pratt, Martin Price, Bob Rhoades, Sanga Sabhasri, Fausto Sarmiento, Daniel Smiley, Peter Stone, Brad Snyder, Juha Uitto, Pisit Voraurai, Shi Yafeng, and Rimma Zimina.

Funding for the extensive fieldwork, international travel, organization of conferences, and publication of results was contributed primarily by the United Nations University. This was substantially augmented by UNESCO, UNEP, IGU, the Swiss government (SDC), the Chinese Academy of Sciences, the Soviet and Russian Academy of Sciences, the Tajik Academy of Sciences, Yunnan Academy of Social Sciences, the Nepal National Planning Commission, the U.S. National Science Foundation, the (Canadian) International Development Research Centre, the Ford Foundation, Mohonk Preserve, and the International Centre for Integrated Mountain Development.

I am personally indebted for the award of a John Simon Guggenheim Memorial Fellowship, for a sabbatical year at the Geographical Institute, Bern University, 1976–1977. The fellowship and my location in Switzerland at the time Walther Manshard was on the point of taking up his appointment as vice-rector of the United Nations University opened the way for the most rewarding section of the mountain journey.

Versions of the manuscript were read by several colleagues. I am greatly indebted for many valuable comments, criticisms, and general advice. All errors or omissions, of course, are solely my responsibility. The entire text was read by Roger Barry, Axel Borsdorf, and Bruno Messerli. Walther Manshard read Chapters 1 through 10; Hans Kienholz read Chapter 10; Seth Sicroff made a valuable contribution to Chapter 14; Don Alford and Jörg Hanisch read Chapter 15; and Chapter 16 was critiqued by Don Alford, Alton Byers, Jörg Hanisch, and Teiji Watanabe.

David Harrison, in Vancouver, provided vital assistance as overall consulting editor; he smoothed out my linguistic distortions and consistently helped me find the right pathway through my periodic writer's blocks. Dave Andrews, of Digital Art and Restoration, Ottawa, undertook the meticulous preparation of the photographs and drafted the line drawings. I am indebted to the International Centre for Integrated Mountain Development (ICIMOD) and several colleagues for permission to use photographs identified individually in the relevant captions. Bina Gajurel Mainali undertook the task of producing the index. Kumar Mainali created the book. Kumar's dedication and precision has astonished me. Debendra Karki coordinated the task of its printing in Kathmandu.

Perhaps my most heartfelt thanks are owed to the many mountain people I encountered along the mountain journey. They not only tolerated intrusions into their homes and working lives, they provided exceptional hospitality, advice, and sympathetic understanding. For decades, they have been the victims of "developed world" misunderstanding, even scorn and ridicule. While the harsher elements are rapidly becoming a thing of the past, the so-called advanced societies still have much to learn from them. For me, it was one of the great privileges and joys of life to be welcomed into their many differing worlds. I only hope that the pronounced effort now being undertaken in partnership with them by a very large number of individuals and institutions will eventually lead to actual sustainable mountain development.

My final words of thanks are to Pauline Ives and our children, Nadine, Tony, Colin, and Peter. Pauline not only kept *Mountain Research and Development* on track for nearly two decades; she kept the entire family on track at all times. I owe thanks to them all for tolerating my long absences and for still listening with enthusiasm to the many improbable stories I brought home with me.

Ottawa, Canada
September 2012.

Fig. 1. The Torngat Mountains, northern Labrador, Canada, with Chasm Lake forming a diagonal across the centre ground. Location of some of the first fieldwork based on the McGill University and Arctic Institute of North America graduate student scholarship programme, 1955 to 1960.

CHAPTER 1

Global Mountain Awareness

Mountainous landscapes occupy about a fifth of Earth's entire terrestrial surface, and their inhabitants number more than a tenth of humankind. Until as recently as the 1970s, mountains remained far beyond the reach of mainstream world affairs. Except for a few regions, such as the European Alps, Northwest Europe, and the North American western cordillera, they were regarded as remote, harsh, inaccessible environments of little interest to anyone other than mountaineers, explorers, and scientists. Trekking tourism had barely entered its formative stage. Furthermore, the mountain peoples who had lived in these environments for generations had been largely ignored.

By 1992, however, during the Rio de Janeiro Earth Summit (UNCED), mountain regions had begun to attract worldwide attention. The preceding twenty years, beginning with the United Nations Conference on the Human Environment, held in Stockholm in 1972, had seen the emergence of concern for the world's environment, leading to the first clear definition of "sustainable development." While mountains had remained a poor relation to this rapid growth in environmental concern and urgent calls for action, the small group of mountain advocates with which I was associated was provided an opportunity to infiltrate the environmental power structure. This culminated during the 1992 Rio Earth Summit when mountains were accorded comparable standing with other major problems of international scope, such as the tropical rain forests or the oceans. As part of this undertaking, I was privileged to act as an official representative of the United Nations University (UNU).

After Rio, with an almost unseemly rush, mountains were recognized as "the world's water towers" and came to be perceived as vital to global sustainability. Indeed, the United Nations Organization (UNO) declared 2002 as the International Year of Mountains and December 11 as the annual International Mountain Day. The year 2012 (twenty years after Rio) was one of reassessment, serious introspection, and renewed calls for greater efforts on the part of all the UN member states and many agencies, institutions, and individuals. In June 2012, there was a major reconvening in Rio de Janeiro in the form of the United Nations Conference on Sustainable

J. D. Ives, *Sustainable Mountain Development*, https://doi.org/10.1007/978-3-030-96029-2_1

Development (UNCSD, or Rio+20). Nonetheless, to what extent mountains will have retained a commanding position of concern in the world community unfortunately remains in doubt at this time.

How did all this come to pass? What provoked alarm among conservationists in the early 1970s and led to urgent demands for action? Suddenly, the Himalaya, for instance, were perceived to be on the edge of environmental catastrophe. The World Bank in 1979 predicted that as Nepal had lost half its forest cover within a thirty-year period (1950–1980), by A.D. 2000 no accessible forest would remain. The executive director of the United Nations Environment Programme (UNEP) was quoted as saying in Dhaka in 1990 that "chronic deforestation in the Himalayan watersheds was already complicating and compounding seasonal floods in Bangladesh."[1] The World Resources Institute in 1985 claimed that "a few million subsistence hill farmers are undermining the life support of several hundred million people in the plains."[2] All three of those statements, made with such international authority, carried remarkable political implications.

Massive deforestation, primarily by poor mountain farmers, was seen as an unstoppable process that would strip the entire 3000 kilometres of the mountain range of its protective forest cover by the end of the twentieth century. This in turn was projected to cause massive soil erosion, innumerable landslides on the steep unprotected slopes, and extensive flooding and siltation across the flood plains of the Ganges and Brahmaputra all the way down to the Bay of Bengal. Hundreds of millions of inhabitants of northern India, Bangladesh, and Pakistan would be subjected to ruin, starvation, and death. Political tensions would be seriously heightened.

This book is an attempt to explain, at least partially, these significant issues and how, along with what was initially a small group of academics, I was drawn into an expanding international political process. I endeavour to trace this process as a personal journey that describes serious and sustained teamwork and multi-institutional collaboration along with tales of comradeship, privileged contact with mountain people, and high adventure. I will portray the mountain people, who welcomed me into their homes, as repositories of great wisdom, sources of often unlimited hospitality, exemplars of persistence and ingenuity, and yet victims of oppression and neglect, frequently characterized as "ignorant peasants" who were "destroying their own environments."

From this earliest awakening to worldwide political and environmental mountain controversy, I was the beneficiary of being part of a small group of academics who were brought together through the International Geographical Union (IGU), the UN Environmental, Scientific, and Cultural Organization (UNESCO), and especially the United Nations University, to focus attention on mountain problems.

The first step in our task, the magnitude of which we could not remotely conceive at the beginning, was to use our academic training, and especially our professional scepticism, to undertake field research to test the authoritarian and apocalyptic assumptions that were being proliferated worldwide. The second step was to challenge them on the basis of our accumulating scientific evidence. And third, we had to attempt to eliminate the false speculation and replace it with a measured, research-based common sense. We realized that, to achieve this, we would first need a broad international and interdisciplinary research effort. We would then have to penetrate the international political apparatus of the United Nations Organization itself. This would mean contesting a number of cherished convictions held by powerful authorities, which we came to call "mountain myths."

The 1992 Rio Earth Summit and the 2002 International Year of Mountains, while far beyond our contemporary horizon in the early stages of our growing awareness, eventually became our objectives. However, to achieve the necessary access to the top level institutions, we would need to convince top level individuals on the world stage to guide us through the intricate pathways of international bureaucracy and intrigue. So it was that I began my personal journey towards the critical 1992 Rio conference and eventually to UN headquarters in New York in December 2001, where I would deliver a keynote address on behalf of UNU Rector Hans J.A. van Ginkel.

My Fascination with Mountains

As I travelled to New York, I reflected that this whole enterprise had unfolded over many years since the 1940s, from a boyhood fascination with mountains and high latitudes, initially prompted by two early fishing trawler voyages to Svalbard and the Barents Sea. I grew up in Grimsby, England, in those days one of the world's largest fishing ports. My home town, situated in northeast Lincolnshire, presented me with great expanses of intertidal sand flats and a commanding "high" point in the Lincolnshire Wolds, though little more than 100 metres above the North Sea. It was hardly surprising, therefore, that my fifteen-year-old's views of the Lofoten Islands, Svalbard's glaciers and ice caps under the midnight sun, and the pack ice off Franz Joseph Land close to 80°N, swept me off my feet. Many years later, when my students would ask me to account for my enthusiasm for mountains, I would jokingly reply that I'd been born too close to high watermark on a gently shelving coast and couldn't swim.

There followed a series of undergraduate expeditions to Iceland's glaciers and ice caps from the University of Nottingham. This, in turn, led to my meeting with Pauline, a University of London geography student who was also intent on adventure in Iceland. At the same time, McGill University had begun developing an ambitious research programme in the Canadian Arctic and Sub-Arctic. In the spring of 1954, as I was camping alone in Iceland, nursing my favourite glacier Morsár-jökull, my local farmer friend Ragnar Stefansson delivered the scant mail on horseback. There was a letter from a Professor F. Kenneth Hare, chair of the McGill University Geography Department, offering me what became a McGill-Carnegie Arctic Research scholarship. This precipitated marriage with Pauline and our arrival in Montreal as proud Canadian landed immigrants the following September.

My first post-doctoral appointment was director of the McGill Sub-Arctic Research Laboratory (1957–1960) in central Labrador-Ungava during the 1957–1958 International Geophysical Year. However, it was participation in the 1964 London International Geographical Congress, as the newly appointed director of the former Canadian Geographical Branch that led to my first contacts with worldwide scholarship and teamwork. There I was introduced to Professor Carl Troll, who was completing his term as IGU president, and to Academician Innokentiy Gerasimov who, as senior vice-president, under normal circumstances, would have succeeded Troll as president.[3]

Carl Troll subsequently invited me to be a member of his newly created IGU Commission on High-Altitude Geoecology in 1968 and to succeed him as its chair in 1972. This required a confirmation vote by representatives of all member countries of the IGU, and Academician Gerasimov surprised me by promising to assert his influence with Soviet satellite and Third World member countries to ensure my election. This represented a critical development that led me into all that follows in this narrative.

From this point, the speed of the journey into global mountain research and political, environmental, and development controversies accelerated. The single most important link that Carl Troll facilitated was my meeting with Professor Bruno Messerli, Bern University, Switzerland. This resulted in a year-long sabbatical on a Guggenheim fellowship as guest professor at Bern University and in a thirty-year partnership and close friendship

that took us both to the Himalaya, the mountains of China, the Soviet Union, and to mountain regions on all continents except Antarctica. It also took us to Rio de Janeiro in 1992 and to UN headquarters in New York in 2001. Throughout most of this period, 1972–2002, we alternated as chair of the IGU mountain commission established by Carl Troll.

To introduce the episodes of high adventure in this personal narrative, I will refer briefly to three of the more notable escapades that are recounted in detail in their respective chapters. The first, because of my political "immunity" assumed by colleagues in the Scott Polar Research Institute, Cambridge, and based on my close collaboration with Academician Gerasimov, I was asked to assist in the extraction of a Russian Jewish glaciologist and his family from the Soviet Union in 1976. The second, in 1979, was an effort to extract myself at knife-point from three armed bandits in the jungles of Sikkim. And the third involved being arrested by Kalashnikov-wielding militia in the Pamir in 1999 while on a mission to determine the hazards posed by a huge landslide-dammed mountain lake. In distant retrospect, these kinds of episodes fade into the category of great yarns to spin to one's grandchildren. At the time, however, as the reader will come to understand, they were rather more serious and, it seems, an unavoidable part of my mountain commitment, however unconventional.

On less dramatic, but much more significant grounds, as the mountain journey unfolds in this volume, I show how we were able to use the growing impact of the mountain commission's association with UNESCO's

Man and the Biosphere (MAB) Programme ("Project 6: Study of the Impact of Human Activities on Mountain Ecosystems"), coordination with UNU's new project on "Highland-Lowland Interactive Systems," and to establish the International Mountain Society in 1980. The Society provided us with our own apparatus for international dissemination of the results of mountain research and the means of expanding our communications network.

Several of the major UN agencies as well as the worldwide news media in the 1970s and 1980s were blatantly forecasting that a Himalayan catastrophe was imminent. In large part, this coloured the formal objectives of our new mountain society: "To strive for a better balance between mountain environment, development of resources, and the well-being of mountain peoples." It also motivated our challenge to the new environmental paradigm of the day that I encapsulated in the term "theory of Himalayan environmental degradation."

These early initiatives led to two international mountain conferences at Mohonk Mountain House in upper New York State (1982, 1986), both largely financed by UNU. They brought together a group of dedicated mountain academics[4] and senior international authorities. They helped establish contact with Maurice Strong who was serving as under secretary-general of the UN and senior advisor to Canadian prime minister Pierre Elliot Trudeau.[5] Maurice agreed to be honorary chair of the second Mohonk conference and provided us with a special entrée to the Rio Earth Summit in his subsequent role

as its secretary-general. The Mohonk conferences led to the publication of *The Himalayan Dilemma: Reconciling Development and Conservation* (Ives and Messerli 1989), which delivered a serious challenge to the theory of Himalayan environmental degradation. It opened up the stimulating discourse and environmental and developmental controversy that came to be identified as "The Mohonk Process."

The core group that set in motion the Mohonk Process constituted themselves informally as the champions of the "Mountain Agenda" and during the following years there was a progressive addition to the membership.[6] The group became known, mischievously amongst our graduate students of the time, as the "mountain mafia." At this point we had firmly entered the political arena and assisted Maurice Strong's efforts to include the "mountain chapter" (later to become Chapter 13) into Agenda 21 at the Rio Earth Summit. Our protagonist group was staunchly encouraged by UNU Vice-Rector Walther Manshard and, politically and financially, by the government of Switzerland.

It will by now be clear to the reader that one of the main themes permeating this book is the need to subject major environmental mountain alarms of the day to careful scientific research, even if it requires a challenge to the presumptions of dominant international authorities.

Another of my themes, of an entirely different nature, is the incredible happenstance, coincidences of meetings and events, and unexpected opportunities to present our views to powerful personalities at critical moments and, in turn, of being influenced by those remarkable individuals. The personalities you will encounter in these accounts include Maurice Strong, Deng Xiaoping, Their Majesties King Bhumibol and Queen Sirikit of Thailand, His Holiness John Paul II, Prince Sadruddin Aga Khan, and Lord John Hunt, leader of the 1953 Mount Everest Expedition that led to Hillary's and Tenzing's first ascent. The academic contacts were equally important: Carl Troll, past president of the IGU and founding chair of the Commission on High-Altitude Geoecology; Walther Manshard, vice-rector of the United Nations University; Jim Harrison, my onetime chief in Ottawa and former assistant director-general of UNESCO; and Frank P. Davidson, "macro-engineer" of MIT and Harvard. This theme of "help in high places" relates both successes and failures associated with being able to contact the "right" people, sometimes at the "right" time, sometimes not. Yet throughout the efforts to transfer the results of our academic research into the political arena, the importance of UNU's status as an academic institution, special amongst other UN bodies, was repeatedly demonstrated.

Chance encounters in various parts of the mountain wilderness throughout the world and disputes during numerous international conferences and consultations led me into many exciting ventures. There were moments when a sense of achievement was justified, yet there were countless others of frustrating failure and disappointment. But there was also the redeeming joy and exhilaration of meeting and working with minority mountain people – in particular, discovering

that their innate wisdom so often put to shame the so-called Western experts' "scientific" superiority that aimed to correct their ways and so improve their living standards. We learned to appreciate that the mountain farmers and herders were frequently not part of the problem; rather, they were essential to forging solutions, but were so often the overlooked victims of the world at large.

Thus emerges another key theme of this book: the realization that we had so much to learn from the mountain people themselves, and that our objective in entering their territory was to ask them for advice and explanation, and to work with them towards better understanding. In today's world, this theme might appear strangely obvious or irrelevant. But thirty to forty years ago, we were continually grappling with a succession of attempts to impose Western scientific "wisdom" and superior technology on "ignorant peasants" who were categorized as ignorant simply because they were poor, very different, and apparently simple in their way of living.

My account, therefore, relates many flaws in the grandiose UN and Western system and in policies of bilateral aid agencies and NGOs. It demonstrates how a few colleagues working together, provided they acquire some measure of success in struggling through the established system, can achieve something worthwhile despite having no permanent institutional base.

The latter part of the narrative turns from our efforts to counter the forecasts of imminent Himalayan disaster through assumed total deforestation and its inevitable dire consequences. It examines the recent rise in a comparably impassioned and equally misguided exaggeration that climate warming is leading to the rapid disappearance of Himalayan glaciers, devastating floods, and the ultimate drying up of major rivers and widespread death following drought.

This is an especially sensitive theme. Climate warming is a generally accepted phenomenon that is probably the single most serious threat to world human and environmental stability, a grave concern that I endorse absolutely. The rapid shrinkage of sea ice in the Arctic Ocean and the thinning and retreat of glaciers over much of the world (even the total disappearance of some small glaciers) are well documented. It is of critical importance not to diminish the sense of threat so posed as worldwide response is urgently needed, yet not to exaggerate circumstances when such exaggerations can be easily exposed.

Nevertheless, the Himalaya, because of extreme altitude, cannot be compared to the Alps or other mountain areas where loss of approximately 50% of glacier volume since 1850 is not in doubt. But as we came to grips with the prediction of Himalayan environmental collapse in the last century, so today there are repeated alarms that all Himalayan glaciers will disappear in the near future; that when they are gone, the diminished supply of water to such vital rivers as the Ganges will reduce them to seasonal streams; and that millions will die from the ensuing drought. In the meantime, there are dire predictions that while the glaciers are retreating and thinning, glacial meltwater lakes are forming and will inevitably burst catastrophically, bringing death and destruction to millions downstream.

The final chapters of this book are thus devoted to an examination of these gross exaggerations, if not falsifications. It would appear that the news media and some of the authoritative international agencies cannot avoid the temptation to spread magnified alarms of imminent catastrophe. I believe that, as with false alarms of disaster through deforestation from 1970 onwards that started me on my personal journey from academia to international politics, so the current waves of exaggeration must be shown for what they are. False alarms can cause enormous damage, in this case, both to mountain peoples and to the mountains themselves.

Fortunately, significant new financial support is beginning to accelerate the international cooperation needed to more fully understand the environments of some of the higher places of our world. Such understanding is all too necessary if the mountains are to play their part in the struggle to maintain worldwide environmental sustainability.

Fig. 2.   The Inugsuin Pinnacles at midnight, northeast coast of
Baffin Island close to 70 degrees north. Geographical Branch,
Canadian federal government expeditions, 1961 to 1967.

CHAPTER 2

An International Mountain Involvement
The Beginnings

It was London, August 1964. The occasion was the International Geographical Congress (IGC). These meetings were geographical extravaganzas held at intervals of four years and usually rotated amongst the major capital cities of the world. I had been appointed director of the Canadian Geographical Branch the previous April. Despite my strong commitment to fieldwork in Baffin Island in the Canadian Arctic, my friendly but stern chief, Dr van Steenburgh, departmental director-general of scientific services, had firmly recommended that I attend the London conference. He insisted that this was necessary to reflect the change in my status from an Arctic field researcher to wider administrative responsibilities. Little did I realize that it would be the first step towards involvement in international mountain geography, political controversy, and adventure.

On entering the main conference hall, I immediately encountered Professor Gordon Manley, my wife Pauline's former University of London professor, who had visited us in both central Labrador-Ungava and Ottawa. He presented me to Lord Dudley Stamp, the pre-eminent British geographer, who, in turn,

introduced me to Professor Carl Troll, the outgoing IGU president. Troll, to me, was a legend. Past middle age, trimly bearded with military posture, I likened him immediately to one of Rommel's World War II panzer tank commanders. This was far from accurate, but I was very impressionable at the time. His Germanic bearing and mannerisms were striking but did not conceal a warm personality. Although very junior, I had two things in common with him: mountains and the cold regions of the world.

My brief conversation with Stamp and Troll was interrupted by the approach of another commanding figure. Academician Innokentiy Gerasimov was not altogether relaxed, although he did shake my hand; he nodded curtly to Troll and appeared about to ignore Stamp except that he (Stamp) was ready to open the general meeting with both Troll and Gerasimov as principal speakers. It was generally assumed that, as senior vice-president, Gerasimov would automatically receive the formal vote of the Union to succeed Troll as president, thereby ensuring that the 1968 Congress would be hosted by Moscow.

I took a seat with Gordon Manley close

© The Author(s), under exclusive license to Springer Nature Switzerland AG 2022
J. D. Ives, *Sustainable Mountain Development*, https://doi.org/10.1007/978-3-030-96029-2_2

to the front and awaited Troll's presentation, to be followed by what was assumed would be Gerasimov's lead-up to the presidency for the following four years and the formal invitation to Moscow for the 1968 meeting of the IGC. Yet the tension was palpable. My curiosity and excitement were being finely honed, so much so that I immediately forgot the content of Troll's speech. Gerasimov took the podium and instantly launched into a tirade against the British government and British geographers in general. The gasps of surprise from a large international audience were quite audible. Apparently members of the East German delegation had not received their visas in time for entry into the UK. Gerasimov rebuked the British establishment for playing Cold War politics and resigned as vice-president, thereby throwing this opening meeting into confusion. He quit the podium and, with the other members of the Soviet delegation (and presumably their KGB minders), left the meeting. Troll and Stamp were faced with the delicate task of trying to smooth things over. And – surprise! The new president was Professor S.P. Chatterjee and the new location for the 1968 Congress, New Delhi.

What is now so remarkable to me about this, my first pitch into geographical and academic international affairs was that, subsequently and far beyond my ken at the time, both Carl Troll and Innokentiy Gerasimov came to play significant roles in my later mountain endeavours. Equally far beyond my remotest imagining, many years later, in November 1976, I was invited to present a eulogy to Carl Troll at the celebration of his life in the cathedral at Bonn. Many more years later, I

paid similar homage in Moscow to Academician Gerasimov.

I duly attended the 1968 Congress in New Delhi but in very different circumstances from those of London in 1964. Much had happened in the four-year interval. My Baffin Island commitment, involving field parties exceeding 30 personnel with fixed-wing aircraft, helicopter, and icebreaker support had been very successful (Ives 2014). Nevertheless, the vicissitudes of Canadian government in-house politics in 1966 and 1967 had led to the disestablishment of the Geographical Branch. I had been offered a senior appointment in the prestigious Geological Survey of Canada. Although I was tempted to remain in government service, I ardently considered myself as a geographer with a strong attachment to fieldwork.

Thus, by late 1967, I was ensconced in Boulder, Colorado as director of the University of Colorado's Institute of Arctic and Alpine Research (INSTAAR) with an academic appointment as professor of geography. The dean of the university graduate school, keen to attract me to Colorado with my family, had facilitated university appointments for several of my key Baffin Island colleagues as faculty and support staff. He also agreed to support my participation in the New Delhi Congress, together with an extensive tour of Asian mountain areas, including Japan, the Darjeeling Himalaya, Afghanistan, and Uzbekistan.

International geographical congresses were opportunities for presentation of the results of recent research in many of geography's sub-disciplines as well as for the formal business meetings of the IGU. They included a series of field excursions to demonstrate the geography and associated research and applied problems of the host country. Finally, the various IGU commissions were able to hold their own specialized meetings, often accompanied by field excursions; so, as a physical geographer, or geomorphologist,[7] I was closely associated with a commission that had planned a field meeting based on Darjeeling, a renowned gateway to the Himalaya.

The Darjeeling meeting, which preceded the main congress in New Delhi, had been planned for a group of 25 mountain enthusiasts hosted by the University of West Bengal. My participation had a considerable impact on my thinking, leading to my earliest doubts about the conventional assumption that massive deforestation by "ignorant" mountain farmers would result in environmental devastation in the Himalaya. The serious practical and political issues surrounding this worldwide assumption somewhat later became a major involvement for me and constitutes one of the prime themes of this book.

The main Delhi Congress was significant in the short term.[8] As part of the main meetings in New Delhi, I had been invited to chair a special session of the geomorphology group. Carl Troll entered the lecture room as I was bringing the meeting to a close. He suggested a private coffee break during which he enquired of my activities over the previous four years. He explained that his proposal for creation of a new commission on high-altitude geoecology would be formally approved. As commission chair, he would need to form a small international secretariat to run it. Consequentially, I would find a written invitation awaiting my return to my university office in Colorado. We parted with highly formal handshakes and the full understanding that we would be working together in the near future. The Darjeeling symposium and my second meeting with Troll, this time in New Delhi, led to involvements that I could never have envisioned at the time.

The early years in Boulder, Colorado, as director of the university's Institute of Arctic and Alpine Research (INSTAAR) proved both challenging and enjoyable. The first task was the incorporation of former Baffin Island and Canadian Geographical Branch colleagues John T. Andrews, Roger G. Barry, Nel T. Caine, Patrick J. Webber (all professorial appointments) and Rolf Kihl and Kathleen Blackie (Salzberg)[9] as support staff. This was an essential part of the expansion of a somewhat moribund institute and its potential jewel, the mountain research station, situated at 3000 metres above sea level in the Colorado Front Range. New research funds supported a major undertaking on the alpine ridge above the research station and initiation of the United States alpine section of the Tundra Biome of the International Biological Programme, a large multi-year and interdisciplinary research project in the San Juan Mountains, and an ambitious avalanche research operation in the same area. There were many smaller projects and a series of senior high school and university undergraduate field training projects supported by the National Science Foundation.

By early 1969, I was able to launch a new scientific quarterly, *Arctic and Alpine Research*.[10] In this manner INSTAAR, 18 years after its founding, finally became an internationally recognized university arctic and mountain research and teaching organization. In effect, my position as director of the institute became an important springboard for the initial invitations to participate in international mountain activity.

Carl Troll's appointment of me to his secretariat was apparently only of modest significance.[11] I attended its first formal international meeting to coincide with the International Geographical Congress scheduled for Montreal in August 1972 (see Ives 2014). Troll was insistent, however, that *his* commission was field- and mountain-oriented, so our meeting place would be on the Calgary campus (more than 3000 km west of Montreal), then part of the University of Alberta. Calgary would provide easy access into the Canadian Rockies. Because of this, we needed a special local organizer – Professor Stuart Harris of the Calgary faculty. The first concerns faced by Stuart arose from unexpected problems due to Troll incurring a series of heart attacks during the 1971/72 winter. This resulted in two developments. Troll resigned as commission chair; the first I heard of this was when I opened an airmail envelope from Bonn. It contained the copy of a letter Troll had addressed to Professor Chauncy Harris, secretary-treasurer of the IGU, explaining his reason for resigning and stipulating that "Professor Jack D. Ives *will* [my emphasis] succeed me as chairman." This displayed yet another facet of Troll's approach to life, especially in view of the fact that these appointments were elected ones before the entire IGU (in this instance, the election process would be in Montreal, which

Fig. 3.   Carl Troll in the patterned ground tundra of the Alberta Rocky Mountains during the International Geographical Union meeting of the Commission on High-Altitude Geoecology (1972).

I was unable to attend).

The second development was an anxious telephone call from Stuart: "Jack, what should I do, as I suspect that Troll will want to take part in the mountain excursions? What if this brings on another and more serious heart attack? Can you help me persuade him not to participate?" (Troll had been a very physically active mountain geographer, having spent four years travelling through the Andes and

Fig. 4. The Niwot Ridge alpine research area in the Colorado Front Range above INSTAAR's Mountain Research Station was declared a UNESCO Biosphere Reserve in 1979 by President Jimmy Carter. The photograph shows Barry Fahey in 1971 demonstrating "the bedstead" used for measuring point-by-point surface uplift due to ground freezing in the Fall and collapse during the following Spring thaw.

East Africa in the 1920s and participating in one of the pre–World War II German expeditions to Nanga Parbat). My answer was simple: "Stuart, I'm afraid it is not a case of persuading him not to participate, but to make arrangements for rapid evacuation if he does have a heart attack while ascending a steep slope, because you will never be able to stop him." Well, as it happened, the culmination of Stuart's very real anxiety was to be played out on a steep mountain slope in Banff National Park the following July.

I was very fit in those days. I was able to keep up with a tough group of Canadian graduate students until, after a steep, hot climb of some 500 metres we came to the upper tree limit and sat down to rest. A short while after, the second group of our party arrived, panting and perspiring. In their midst was Professor Troll. He was dressed in a suit, with neck tie and trilby hat – very smart. He calmly walked over to me and, standing tall, exclaimed, "Yack, an old Old Vorld geographer comes to timberline, vunderbar!" It was indeed wonderful yet sad that only a few years later I recited this experience, imitating his accent as best I could, as the closing remarks of my eulogy in the solemn atmosphere of the cathedral in Bonn. As my words faded, Frau Troll came forward from the front row of the pews and embraced me with tears in her eyes. She thanked me for providing a human touch which she was sure Carl would have loved.

The IGU mountain commission activities in Alberta took on a life of their own. Here I became acquainted with Professor Bruno Messerli of Bern University, a relationship that evolved into the most fundamental friendship and collaboration of my mountain involvement. And the shadow of Academician Gerasimov, ever-present at that period, re-emerged in Calgary. Gerasimov did not attend the mountain commission meetings, but came to accompany his wife, Rimma Zimina, one of our commission members, to the main congress in Montreal. He took me aside and passed on the assurance that my commission chairmanship would be confirmed in Montreal as he would see to it that the entire Eastern Bloc and many delegates representing Third World[12] countries would be "told" to vote for me. While this was somewhat disquieting for me at the time, I also developed a long and fruitful collaboration with his wife Rimma, as will be related below. It transpired that Gerasimov had a private plan of his own. Nevertheless, at the time, I had no idea of the full significance of Troll's anointing me as his successor.

UNESCO Man and the Biosphere Mountain Project (MAB-6)

The implications of my succession as chair of the IGU mountain commission became evident during the early winter of 1972–73. One November morning, while at my desk at INSTAAR in Boulder, I received a surprise telephone call. The caller identified himself as Dr Francesco di Castri, secretary of UNESCO's new Man and the Biosphere (MAB) Programme created following the first United Nations Conference on the Human Environment in Stockholm in 1972. There were to be 14 specific projects and di Castri was in the act of setting up Project 6 (a study of the impact of human activities on mountain ecosystems). He invited me to attend the first meeting of MAB-6 as a guest of UNESCO. It would be held in Salzburg, Austria, in January–February 1973. The five-day meeting would include a field excursion to Obergurgl in the Austrian Tyrol.[13]

I immediately accepted the invitation and thereby set on course the next major development of my career. Following the discussion with di Castri, I extended my European visit to develop additional contacts relevant to my responsibilities as director of INSTAAR. I visited professors Tranquillini, Gams, Fliri and Hoinkes in Innsbruck; Professor Fritz Müller and Dr Hans Röthlisberger in Zurich; and Dr Marcel de Quervain at the Swiss Federal Institute for Snow and Avalanche Research in Davos. A visit to the Jungfraujoch research station in the Bernese Oberland was also arranged by Dr Röthlisberger.

The MAB-6 meeting in Salzburg and Obergurgl introduced me to many facets of international science organization. UNESCO had the essential task of recruiting competent scholars to provide oversight and advisory capacity for new programme development. At the same time, as these were essentially international undertakings, every attempt was made to ensure representation from as many of the countries, worldwide, that might be interested. Hence, international committees (and 'panels of experts' as this one was designated) had to reflect West–East–Third World parity, insofar as that could be achieved. Di Castri was an ambitious driving force and was ably assisted by a recent young recruit

to UNESCO, Dr Gisbert Glaser, both operating under the eagle eye of Maurice Batisse, director of UNESCO's Division of Ecology and Earth Sciences. Unbeknown to me at the time, both Batisse and di Castri reported to Professor Walther Manshard, who was director of UNESCO's Division of Environmental Sciences. There were a total of 20 members of the MAB-6 panel representing several disciplines, together with a small group of observers from international organizations, such as FAO, IUCN, and ICSU (Appendix 1).

Several individuals came to play a very influential role as an unofficial "core" group: Paul Baker, Corneille Jest, Bruno Messerli, Walter Moser, Makoto Numata, Paul Veyret, and myself. Francesco di Castri asked me to assist with drafting the final report. As I had a very fast pen and a strong sense of opportunism, this led me into a heavy workload because di Castri wanted to publish the final report as soon as possible.[14]

At the Salzburg meeting, we had a distinguished guest speaker, Konrad Lorenz, author of the well-known book *On Aggression,* who was to receive a Nobel Prize later that year. Several of us had the privilege of sharing a suite of rooms with him in a most comfortable *pension*. This led to fascinating conversations over breakfasts and late evening "night-caps." And we began a pattern of attending concerts of classical music. In Salzburg, of course, this included a visit to the Mozart House, but also a performance of *The Barber of Seville*, in German – the choice of language being something of a surprise for me. My growing relationship with the UNESCO MAB-6 and Bruno Messerli was to become intimately tied to opera and classical music (a purely incidental process, but the type of reinforcement that strengthened our commitment to the significant level of work that would be required).

The Man and the Biosphere Programme of UNESCO, with fourteen ambitious projects, laid the foundations for many critically important environmental and developmental activities worldwide. The international designation of biosphere reserves sprang from it (as MAB-8) and an extensive series of developmental and nature protectionist issues received scientific and popular attention; these included rainforests, grasslands, deserts, coastal regions, and wetlands, as well as mountains. Of special importance was the urgent need to bridge the gap between the natural and human sciences and concerted efforts to develop a systems approach and to experiment with computer modelling.[15] In retrospect, MAB was propelled by the 1972 Stockholm Conference on the Human Environment and lessons learned from the International Biological Programme (IBP).

MAB was, and still is, such a vast endeavour that no attempt will be made here to describe it beyond MAB-6. Except for the fact that MAB was essentially a world-scale review of problems facing all earth's ecosystems, it is unlikely that mountains would have been included at all. There was initially strong argument that a mountain project was unnecessary as sections of it would be covered by other ecosystem projects (at a later date, we had to contest the same argument against the "mountain chapter" for Agenda 21 in preparations for the Rio Earth Summit of 1992). Furthermore, as we observed in Chapter 1, mountain regions in the early 1970s, with a number of exceptions, were considered so remote, inaccessible, and of little interest to the general sweep of world affairs that they did not warrant expenditure of intellectual and financial effort. No concern was expressed about mountains during the 1972 Stockholm Conference, despite having Maurice Strong as its secretary-general (Strong subsequently became the single most effective high-level mountain advocate). This virtual "anti-mountain" atmosphere prevailed in many of the corridors of intellectual power in the early 1970s and subsequently became a critical concern of our expanding group of mountain activists.

Nevertheless, following the Stockholm Conference, the UNESCO MAB initiative received a remarkable boost. The United Nations Environment Programme[16] was created with Maurice Strong as founding secretary-general; numerous national governments established departments of the environment, frequently at full ministerial level. It is worth posing the question again at this point: How did mountains reach centre-stage on world affairs with the United Nations 2002 designation of the International Year of Mountains?

Prior to 1972, mountains had attracted only scattered academic interest, based largely on the commitment of individuals or small groups, often of mountaineers. Mountaineering groups and individual academics in France, Switzerland, Germany, Austria, Italy, the United Kingdom, and the United States, had established a long tradition of mountain research and exploration. Following World War II, the International Union

Fig. 5.　Obergurgl and the Austrian Alps. From the summit of Hohe Nebelkogel at sunrise in February 1977. After a night shared with Dr Walter Moser, who introduced the 'Obergurgl Model' during the first meeting of the UNESCO Man and the Biosphere, Project 6: Study of the impact of human activities on mountain ecosystems (Salzburg, 1973).

for the Conservation of Nature and Natural Resources (IUCN) developed an interest in mountains, and the initiation of the human physiology section of the International Biological Programme (Man in the Andes) ensured U.S. scientist Paul Baker's involvement in the MAB-6 process. Equally far afield, Corneille Jest of CNRS, France, developed an outstanding programme in the Himalaya, also focussing on anthropology and, initially, Tibetology. There was also pioneering work in Norway and Sweden dating from the close of the nineteenth century. But until the formulation of MAB-6, there was no international coordination for the mountain cause, with the exception of the newly established, but grossly underfunded, mountain commission of the IGU.

The wide scattering of mountain experience provided UNESCO's MAB-6 initiative with a range of possible participants for its Salzburg panel of experts. Carl Troll's creation of the IGU Commission on High-Altitude Geoecology was a timely beginning of international academic cooperation.[17] His early retirement as commission chair was an occasion to link with MAB-6 and change its name to "Mountain Geoecology." This provided the opportunity for deliberate widening of its mission. Thus, its overall objective became the study of entire mountain ranges and not just the area from the upper treeline to the summits that constituted Troll's original objective. The inclusion of populated mountain areas and related human problems brought the IGU commission into close alignment to the objectives of MAB-6.[18] The change of name and mandate was formally approved during the Moscow Congress in 1976 (Chapter 5). This provided the base for expansion beyond the academic arena and into a study of the problems facing impoverished mountain farmers and environmental pressures, initially in the Himalaya and northern Thailand, and in the Andes. The name change was one of the essential moves leading into several of the main themes related in this book.

The Salzburg panel of experts became the initial building block. However, UNESCO encouraged the panel members to aim for a very powerful goal – to define applied research plans for international and interdisciplinary cooperation for the Himalaya, the Andes, the mountains of East Africa, and the Alps. These locations, of course, were all alluring magnets for many of us who were imbued with a dream of combining "useful," as well as intellectually stimulating, research with travel and adventure in some of the more exotic and inaccessible regions of the world. This was especially relevant as I believe it led to the long-term commitment of many academics who were primarily interested in the possibilities of innovative research and were not necessarily seeking to augment their academic salaries – the involvements were strictly *pro bono*!

Before we left Salzburg, di Castri had tempted us to continue work on defining objectives with the promise of a series of UNESCO-financed meetings in Lillehammer (Norway), Vienna, Washington, La Paz, Kathmandu, Bogotá, Ruanda, and Briançonnais (France). Details of the more vital of these meetings are provided in later chapters in chronological sequence.

Within weeks of my return home to

Colorado, di Castri telephoned to inform me that the Norwegian government had agreed to host the next MAB-6 meeting in Lillehammer, located 150 km north of Oslo, the following November. But this time the designation "panel of experts" had been upgraded to "international working group." One of the organizational aims would be to set up a formal and continuing international working group. He asked if I would let my name stand for election as its chair.[19]

Lillehammer and Formulation of an International Mountain Programme

The MAB-6 Lillehammer meeting was held in a dark and rain-swept Norway, 20–23 November 1973. It was one of the most frenetic and exhausting meetings I had ever experienced. I was elected chair of the newly established international group that had been enlarged to embrace both mountain and tundra ecosystems; this mirrored the name of the institution which I served as director (INSTAAR).

My first delicate task as chair was to handle an immediate demand to hold the meeting to the international bilingual standards of the time. There were only two French participants, both of whom spoke good English, but most of the others had little or no French. The ensuing need for running translation would be an obvious and serious drag on the proceedings. After my "election" as chair and the standard opening courtesies, we adjourned for the traditional long coffee break for getting acquainted. I mildly admonished

my colleague, Professor Olivier Dollfus – why had he insisted on the bilingual rule? He gave an eminently reasonable answer and a rueful personal apology that was hardly necessary: had he not insisted, his future funding for international meetings would be in danger of cancellation by the French authorities! So we struggled on.[20]

The next note of discord sprung from a special early presentation by Dr Fred Bunnell, University of British Columbia, who had been invited by UNESCO as an observer. His task was to introduce, with examples, the need for adoption into future prospective MAB-6 research of a systems approach. There was so much dissent, especially from many of the Third World participants who, at that time, had barely heard of computer programming and systems analysis, that I felt obliged to temporarily ban from the proceedings any use of the terms "systems analysis" and "computer modelling." Nevertheless, the justification for applying the latest in research technology was generally recognized. It did play a major part in subsequent developments, and my apparently strange behaviour was a very temporary expedient. This introduction of computer technology in 1973 represented an early example of the vastly enlarged approach currently used, for instance, in climate change research.

After these two quibbles, the work at Lillehammer moved into high gear, and it was the single most exhilarating and productive meeting of the entire MAB-6 series. Several of the original Salzburg panel were present: Paul Baker, Corneille Jest, Walter Moser, and Makoto Numata providing a strong sense of continuity. The UNESCO MAB secretariat had been expanded to include Malcolm Hadley, who was fortunately an excellent writer and editor as well as a thoroughly engaging colleague.

A series of documents were drafted that were eventually amalgamated into the final report under the following headings:

1. Approaches to integrated mountain studies
2. Problems of resource development and human settlement in tropical mountain regions
3. Problems of tourism, technology, and land-use alternatives in temperate mountains
4. Problems of land use in high-latitude mountain and tundra ecosystems with special reference to grazing, industrial development, and recreation
5. Comparative worldwide research

There were eight substantial appendices, three of which proved vital to what subsequently gave a great spurt to future mountain research:

- Initial steps in planning regional collaborative projects in the Himalaya;
- Statement of the urgency to examine similar problems for the Andes; and
- Summary report of a regional MAB-6 meeting between countries of the European Alps.[21]

The effort that went into the Lillehammer document, published in March 1974, was phenomenal, and I still believe it was fundamental to the entire international mountain applied research process that evolved subsequently. It led to a series of influential regional meetings in Vienna, Washington, La Paz, Bogotá, Kathmandu, Boulder, Briançonnais, and probably most significantly to a combined UNESCO– German Foundation for International Development (GTZ) meeting in Munich in December 1974 (Chapter 3).

Vienna 1973 and MAB-6 Planning Conferences

The December 1973 meeting, hosted by the City of Vienna and the Government of Austria, was the first in the series of regional conferences that detailed plans and reinforced existing relevant activities in mountain regions worldwide. Political intrigues also surfaced in Austria, reflecting competition between Innsbruck and Vienna for designation of the prime field area for Austrian MAB-6 research and the assumed related flow of financial resources. The meeting was also set amidst the Viennese festival of St Nikolaus and included a sumptuous dinner hosted by the Bürgermeister, complete with Romani violins and dancers.[22]

Recognition of the beginnings of what evolved as the "old guard" of academic mountain activists surfaced in Vienna: thus, Paul Baker, Corneille Jest, Bruno Messerli, Walter Moser, and I found common cause leading to long-term mutual support. The final Vienna MAB-6 social event was an evening at the Staatsoper – a superb performance of *Madam Butterfly* and yet another reinforcing experience.

After the Vienna discussions, Bruno had invited me to visit Bern and his family. Bruno introduced me to his wife, Beatrice, in the Münsterplaz after we had witnessed sunset on the Oberland from the main tower. Not to be outdone by the Austrians, they then took me into the Münster for a performance of Handel's *Israel in Egypt*. I was very surprised to see that the title of this timeless masterpiece had been changed to simply *Israel*: the Israeli army was indeed deep into Egypt and political sensitivity ruled the day, and we entered the cathedral through an impressive security cordon. Afterwards, despite my red hair, rather than the traditional dark colouring of the saint, I was introduced to the four young Messerli children as the graciously heralded St Nikolaus stranger whose duty was to hand out a series of presents (provided, of course, by their parents). Salzburg and Vienna established a tradition for Bruno and myself – a combination of our mountain commitment with opera and classical music.[23] Indeed, this pattern of cultural activity extended to include indigenous folk music and cultural events; it continued throughout the mountain journey, greatly reinforcing our sense of linkage between the local people we met and worked with and the mountains in which they lived.

La Paz 1974 and an Issue of Citizenship

UNESCO's next regional meeting was scheduled for La Paz, Bolivia. Kathmandu was to follow. These cities were central points for their adjacent mountain ranges; they were also the most expedient, being located in politically neutral (or near-neutral) countries. UNESCO wanted to actively involve the various "mountain countries" rather than just individual "experts." Individuals invited to attend henceforth required formal designation by their national governments. The newly formed U.S. national committee for MAB, for instance, was set to operate under the aegis of the Department of State, so that its chair was a fairly senior department official, namely, Dr Donald King. The proposed La Paz meeting was therefore on the agenda of the March 1974 meeting of the national committee in Washington, DC.

This presented me with a quandary. I was a Canadian citizen permanently resident in the United States with a long-term intention of returning to Canada. As I was to learn, Francesco di Castri had applied another behind-the-scenes piece of leverage to influence my selection for the La Paz and future MAB-6 meetings. Immediately before the Washington meeting, therefore, I was invited into Don King's office for what turned out to be a very private tête-à-tête. Don opened by explaining that questions had been raised about my citizenship – I supposed that prospects for a whole series of expenses-paid travel to a variety of exotic places was quite tempting. But as I was very enthusiastic to continue my role in MAB-6, despite an ardent sense of Canadian nationalism, I indicated to Don that I was eligible to apply for U.S. citizenship as I had completed the necessary five years of residence. Don nevertheless raised a note of alarm: "Whatever you do, that is not the way. It would take far too long to obtain the necessary security clearance for you." He handed me a file and confessed that the decision had already been made: that he had invited me into his office to let me know, circumspectly, who were my "enemies." I scanned the file and was surprised to recognize the signatures of several colleagues who had argued that only U.S. citizens should (could) be considered – themselves? The *quid pro quo* with UNESCO is that the distinguished Paris authority paid my expenses while I sat behind the Stars and Stripes at the conference table in La Paz.[24]

Shortly after returning home from the Washington meeting, I received a phone call from Don King to explain that the La Paz meeting was confirmed. He assumed that I was free to attend as a United States representative, along with Paul Baker. When I confirmed, he asked whether (as UNESCO would cover my expenses and he also had budgeted for me) I would like to nominate a colleague, bearing in mind that Francesco di Castri was expecting a significant contribution from me for report preparation. Then I startled both myself and Don with an impromptu suggestion that I would prefer to have my secretary, Ann Stites, there who could really assist with report preparation, than bring an academic. Don concurred, although I replied that I had better ask Ann first. She didn't need any persuasion.

I still had one anxiety about the trip to Bolivia. During May 1974, despite several years of heavy fieldwork above 3000 metres, I had experienced a mild form of altitude sickness, and the medical prescription I was taking would expire shortly after my arrival in La Paz. I talked with Paul Baker about it, and he assured me that if there was any recurrence, at least I would be in the best possible company as our La Paz meeting was dovetailed with a meeting of the world's leading high-altitude medical experts. So I went ahead with plans to arrive three days early, partly to acclimatize and partly so that I could visit the physiological research station in the Cordillera Real above La Paz.

Andean Mountain Issues

The Andes is the longest north–south mountain chain in the world. It stretches from latitude 10°N for more than 7000 km to latitude 55°S. Proximity to the Pacific Ocean, great altitude, and complex topography ensure an extensive range of climates and vegetation types from equatorial rainforest on the lower east-facing slopes above the Amazon basin, to extremely arid west-facing slopes in southern Peru and northern Chile. In the far south, the Andes reach into the sub-Antarctic with impressive ice caps and glaciers that flow down to Pacific sea level. Initially Alexander von Humboldt, followed by Carl Troll, had characterized the Andes as the world's primary example for studying the influence of a great range of altitude and latitude on natural, or ecosystem, diversity and traditional human adaptations. Physiological response to high altitude hypoxic environments as well as the long period of civilized human history upon which was imposed the Spanish Conquest, followed by modern quasi-colonialism and twentieth century rapid growth in population had resulted in an extremely complex

Fig. 6.   The Bolivian Altiplano looking eastward to the Cordillera
Real. Photograph taken on field excursion during the UNESCO
MAB-6 regional meeting in La Paz (1974).

set of human–nature relationships. The main points that were taken up for discussion in La Paz can be briefly summarized under three regional divisions:

1. **The agriculturally important humid Andes of Venezuela, Colombia, and Ecuador.** The northern Andes were distinguished from the rest of the region by higher relative humidity and great climatic symmetry between their east and west flanks. Emphasized was the need to increase food productivity balanced with environmental stability in each of the agricultural belts, from sea level to the high *páramo* at 3200 m to 4500 m.

2. **The Central Andes of Peru, Bolivia, Northern Chile, and Northern Argentina.** A primary problem here was identified as mass migration of highlanders from the *altiplano*, the great central plateau generally above 4000 m, both to the arid coastal lowlands and to the moist Amazon basin and lower Andean slopes. This region had been the heart of the Inca Empire and much of the traditional subsistence agriculture had survived until well into the twentieth century, although mass movement of poor indigenous peasants into cities of both the dry and moist lowlands was causing serious problems of reverse adaptation to altitude and abrupt life style changes. This process was producing remnant high-altitude settlements deprived by emigration of many of their youthful members.

3. **The Southern Andes,** restricted by the UNESCO mandate to Argentina and Chile north of latitude 40°S. MAB-6 responsibilities would focus on landscape conservation, watershed management, mining impacts, and development of tourism. Here the Andes contracts laterally to a single great ridge with none of the high plateaus characteristic of the Central Andes. With increasing southerly latitude, there is a consequent increase in daily and seasonal temperature amplitude.

The great majority of the participants in La Paz were from the Andean countries, including the observers, who were all from the South American regional offices of FAO, WHO, and UNEP. It was a special pleasure for me to have Paul Baker introduce me to Dr Carlos Monge, founder of the famous Peruvian Instituto de Biologia Andina, whose early identification of chronic mountain sickness (Monge's Disease) had made him a household name amongst mountaineers and mountain scholars.

Overall, the La Paz MAB-6 meeting proved an important step forward from the one held in Lillehammer. Recommendations relevant to the Andes that had been formulated in Lillehammer were unanimously adopted and enlarged upon. In light of the prevailing nature of Andean politics, it is perhaps not surprising that future developments were only partially successful. Nevertheless, the stage was set for international Andean research collaboration. Tentative plans were approved for a series of sub-regional meetings embracing Ecuador–Colombia–Venezuela, Peru–Bolivia, and Argentina–Chile with the next meeting being proposed for Mendoza, Argentina under the chairmanship of Dr Ricardo Luti.[25]

Once again, Francesco was in a hurry – he wanted a draft report completed for presentation to the UNESCO International Coordinating Council meeting, scheduled for

Washington DC the following September. My somewhat unorthodox decision of that time, to arrange for Ann Stites's participation now saw substantial justification as, together with help from Paul, we formulated the various discussions and resolutions into a draft report that Francesco was able to take back with him to Paris for finalizing. One of the unanimously supported recommendations that did bear full fruit in the short term was the production of a state-of-knowledge report on the Andean region.[26]

The field excursions introduced all the participants to some of the major problems facing the Central Andes. We crossed Lake Titicaca to one of the large islands to visit Quechua settlements. A second excursion traversed the Cordillera Real down the precipitous, hair-raising road through the cloud forest to the lower montane forest belts and into the upper reaches of the Amazon rainforest. We also visited an Altiplano llama ranch to witness aspects of life on high and were guests at a splendid llama cook-out.

During the excursions, our very diverse group was able to meet with indigenous mountain farmers and herders and learn of their survival problems on the high plateau. The difficulties created by extensive out-migration to lower altitudes and to environments with which they were totally unfamiliar were reinforced by these first hand contacts.

The La Paz meeting proved a springboard for future applied mountain research in the Andes. It brought together hitherto isolated researchers from the frequently politically fractious Andean countries and established relationships with their European and North American counterparts. Furthermore, while not anticipated in La Paz, when the time came for a concerted push to ensure insertion of the "mountain chapter" (Chapter 13) in Agenda 21 during the 1992 Rio Earth Summit, a network of mountain colleagues was in place, the more effectively to work together.

Shortly after returning to Colorado in late June, I was invited to attend the UNESCO MAB International Coordinating Council meeting in Washington D.C. the following September. While there, I was urged to attend yet another meeting, this time in Munich the following December. In many ways, the Munich meeting was a turning point in the now rapid move towards international mountain cooperation and research.

Fig. 7. Durbar Square, Patan: Taken on the occasion of the UNESCO MAB-6 regional meeting in Kathmandu, 1975, a time of limited vehicular traffic and clean air.

CHAPTER 3

Mountain Strategy in Munich, 1974
On Track to the Himalaya

By 1974, the UNESCO MAB programme was having the anticipated ripple effect. Not only were MAB national committees established in many countries, but the concept of biosphere reserves (the task of another project, MAB-8) was generating a great deal of interest. The increasingly effective alert to mountain problems was also being registered by a number of major unilateral aid organizations such as the German Foundation for International Development (GTZ) and the Swiss Agency for Development and Cooperation (SDC), agencies that had been involved in mountain development problems for some years. In addition, a group of individuals with influence in several powerful organizations had formed an unofficial network that reflected concern about perceived mountain environmental degradation. Together with Klaus Lampe, who shortly was to become director of the Department of Agriculture and Forestry at GTZ, they were the force behind the International Workshop on the Development of Mountain Environment at Munich in December 1974. Many of them were directly familiar with the Himalaya or had acquired strong related interest.[27]

GTZ had recognized the work already accomplished by UNESCO and had obtained the collaboration of Jack Fobes, its assistant director-general. Thus, many of the key MAB-6 collaborators, together with members of the permanent MAB secretariat from Paris, were invited to Munich.

These individuals and more than a dozen other scholars constituted a strong gathering of mountain researchers and administrators, competent to begin a serious discussion on how to tackle the presumed environmental problems facing mountain regions throughout the world, especially in the Himalaya.

The organizing committee had also invited an up-and-coming journalist, Erik C. Eckholm, of the World Watch Institute at Washington DC. Erik's task was to assist the workshop by using his journalistic prowess to attract worldwide attention to the perceived environmental and socio-economic disaster facing the mountains of the world. He succeeded, for the time, to an unbelievable extent. Yet, therein lay the seeds of one of the subsequent problems that came to test our embryo group.

The Munich meeting was the first

© The Author(s), under exclusive license to Springer Nature Switzerland AG 2022
J. D. Ives, *Sustainable Mountain Development*, https://doi.org/10.1007/978-3-030-96029-2_3

occasion when the general concept of "disaster in the Himalaya" was thoroughly elaborated, although more as an informal discussion than as part of the formal proceedings. In essence, the alarm was based upon a series of assumptions: uncontrollable growth of the impoverished Himalayan mountain farming population, resulting in rapid deforestation; steep slopes being cleared for subsistence agriculture and to provide essential fuelwood; and loss of forest cover from steep slopes in a monsoon climate, causing serious soil erosion leading to downstream flooding and siltation. The major impact would be on the heavily populated floodplains of the Ganges and Brahmaputra rivers, causing high mortality and destruction.

Along with a small group of mountain colleagues, I felt a growing concern that much of the assumed alarm rested more on emotion than firm science, a seemingly preposterous deduction in light of the very strong power structure that was arguing the contrary. My concern reflected my experience in the Darjeeling hills in 1968 – again, an instance of assumption without long-term factual support (Chapter 4). Nevertheless, in the presence of so many protagonists, my tentative reaction to the Darjeeling rainstorm left me feeling very uncertain, although still insistent that an extensive data base was required. The dilemma accruing from our doubts about imminent Himalayan environmental and socioeconomic disaster will be reintroduced in later chapters of this book. Yet this dilemma alone heightens the relevance of that meeting in Munich.

The proceedings of the Munich workshop were published by GTZ later in 1975.[28] They contained a series of 13 working papers that had been solicited and prepared prior to the workshop, together with a four-page summary report, a list of recommendations, and a "Munich Mountain Environment Manifesto" somewhat influenced in style by the Club of Rome.[29] It is perhaps remarkable that, although the proceedings and the prior working papers had been thoroughly discussed, they contained nothing that could be classed as sensational. Nor did they place special emphasis on the Himalaya. Most presentations were unconnected discussions of a wide range of problems facing such countries or regions as Ethiopia, Afghanistan, the Central Andes, and the European Alps. They embraced general topics such as mining, tourism, development aid, and the need for alternative institutional structures. There was no formal presentation on the Himalaya.

The overall content of the proceedings clearly emphasized the urgent need for much more applied research and development action. They contained a terse statement to the effect that not only mountains, but also their subjacent lowlands would incur serious consequences, such as downslope flooding, unless a more effective international and collaborative approach emerged. The report contained a series of recommendations, also of a general nature. There was a clear recommendation for establishment of some kind of alternative institutional arrangement and recognition of the logic for close cooperation with UNESCO MAB-6.

This rather modest document, however, was to be almost totally eclipsed by Erik

Eckholm's highly effective journalism. His article in *Science* (Eckholm 1975) followed by his book *Losing Ground* (Eckholm 1976) rank together as one of the most effective and influential alarms of the mountain environmental movement. Eckholm became the most frequently quoted "authority" over the next quarter-century and had a profound effect on academia, the news media, and private citizens. The World Bank and many bilateral aid agencies were influenced by his assessment.

Thus, the first major task of our originally small group of mountain geographers was to demand a stronger scientific base for the alarmist statements that Erik's publications helped to set off. We were faced with a serious dilemma: although we acknowledged that more attention must be given to mountain problems, we were faced with gross exaggeration and misconception. Our struggle is spelled out over the following chapters; the word "dilemma" became the code word for our first major attempt to insist on fact rather than sentiment (Ives and Messerli 1989). In this we were vigorously supported by Michael Thompson, a high-altitude mountaineer and anthropologist, and his collaborators (Thompson et al. 1986).

The workshop itself had moments of drama. There were sub-working groups that generated a great deal of concern and developed much greater force during coffee breaks and along the corridors. The informal Himalayan group certainly contributed significantly to this. However, of particular interest, GTZ (I presumed) perpetrated a highly charged "leak" – to the effect that a very large sum of money would become available for establishment of an international institution devoted to identification and resolution of mountain environmental and developmental problems. This set in motion a level of politicking that greatly impressed me. The European countries within or adjacent to the Alps contained the world's largest cluster of academic institutes specializing in mountain research, as well as the largest group of alpine clubs. The discrete rivalry for a major share of the projected funding began to permeate the corridors of the meeting as it was assumed by the academics present that any such institution would be attached to a Western university. I was even challenged to the effect that it would be inappropriate for me, as director of INSTAAR, an American institute, to enter the competition.

I felt the impact of this so strongly that, at the next general discussion period, I was prompted to risk making a fool of myself. I raised my hand and, upon being given the floor, I explained that, from a purely personal perspective, the selection of a centre of mountain expertise should focus on the area between Grenoble and Vienna. I also indicated that INSTAAR was in no way capable of entering the competition. And in a remarkably prescient manner (at least in retrospect), I called for the establishment of an interdisciplinary quarterly journal that would encourage submission of papers on both mountain research and development issues. My interjection produced some laughter. It also induced Klaus Lampe to take me by the arm and down the street to the nearest beer garden, despite the wintry weather. Since I regard the subsequent encounter as one of those seemingly incidental and minor affairs that, nevertheless, blossom into a remarkable driving force, I will elaborate.

Manoeuvres in a Munich Beer Garden

We had not to walk far. Klaus ushered me inside and sat me down, ordering two beers in a single wave of his hand. As a virtual non-beer-drinker, I sensed this as one of those occasions when I should imbibe. I could not judge whether or not Klaus was angry. It turned out that he was very thoughtful, but by no means angry. He then explained that I had inopportunely brought into the open a controversial issue. There was no intention, he explained, of promoting any European institute. An international institution would be established, but in Tehran, because the Shah had committed a $10 million start-up and, in any case, any such development must be centred on a Third World country. I demurred over Tehran – how would GTZ persuade leading mountain scholars to emigrate to Iran? Klaus asserted that would be no problem. He then changed the topic to applaud my proposal for creation of a quarterly mountain journal but insisted that it should be a university-based enterprise. He then took out his wallet and extracted a 100 Deutsche Mark note. Handing it to me, he advised that all I would need to do would be to persuade several hundred colleagues to contribute a similar amount annually, and I would have a journal. He then cautioned that its editorial board should have French and German members. We returned amiably to the workshop.

The DM 100 banknote from Klaus remained in my wallet for another six years. I like to think that it became the first step in the creation of our quarterly journal *Mountain Research and Development (MRD)*. Founded in 1981, the journal came to play an important role leading to the 1992 Rio Earth Summit and the UN designation of 2002 as the International Year of Mountains. I will revisit later the conviction that Tehran was the ideal location for an international mountain institute.

A number of other important considerations emerged from Munich 1974. Walter Moser presented an expanded version of his Obergurgl Model that included a series of flow diagrams. This was subsequently elaborated by the Swiss and German national MAB committees. Nevertheless, it was the first attempt to demonstrate a method to synthesize history, culture, demographics, the impacts of tourism, and mountain environment (Moser and Moser 1986). An interesting interjection was Frank Davidson's call for a new interdisciplinary approach with his suggestion that we adopt the name "mountainology," thereby taking a leaf from the highly effective creation of oceanography and the outstanding practical performance of Jacques Cousteau. Frank also advised on efforts to put into place a series of institutional arrangements.[30]

One of the primary recommendations that took on a life of its own was for the establishment of at least one new centre for collecting and disseminating scientific and technical information on mountain environments. This initiative was obviously related to my Munich beer garden manoeuvres with Klaus Lampe. I assumed that establishing such a centre was one of the unstated, or understated yet major, objectives of the meeting. My negative reaction to Klaus's belief that the Shah of Iran would be a major player was eclipsed by the ruler's violent overthrow in 1979. However, a radically different approach was being charted well before the Shah's debacle. An international mountain centre was eventually located in Kathmandu, and close contact was maintained with UNESCO MAB-6.

Kathmandu 1975: An International Centre for Mountain Development

The MAB-6 regional meeting in Kathmandu was scheduled for the period 26 September–2 October 1975 with two local excursions. The participants proved an interesting group and many new personal relationships developed. The majority, of course, were from the region that embraced Afghanistan, Bhutan, China, India, Iran, Nepal, and Pakistan (Map 1). However, political dissension quickly reared its ugly head. China was regrettably yet understandably absent; the Sino-Indian border war was too recent and the newly completed China–Nepal highway and Friendship Bridge a decided irritant to India.[31] India had refused to grant visas for potential Chinese participants to travel through its territory, and for them to use the only alternative approach – a road trip from Lhasa through the Himalaya to Kathmandu – was not very practical. I was curious, however, to find no delegation from Bhutan. The only route from Thimphu to Kathmandu at that time was by road through Indian territory, and travel visas were

withheld, something I failed to understand. In light of these decidedly discouraging implications, I was interested to witness the debate between the Indian and Pakistani delegations who would undoubtedly compete for the big prize – the location of a major international institution for applied mountain research and development planning together with significant international funding. National prestige would be involved.

Participants from outside the region hailed from Australia, Austria, France, Japan, USSR, and the USA; they included several of the original MAB-6 working group and key members from the Munich GTZ co-ordinators. The principal organizing group was the UNESCO MAB secretariat from Paris, led by Francesco di Castri. There were also a number of observers representing UN agencies and a large group of Nepalese and Indians. Amongst the Nepalese delegates were two who came to play important roles in subsequent United Nations University involvement in the Himalayan region, Dr Ratna Shumser Rana, who chaired the meeting, and Dr Kamal K. Shrestha, who subsequently became one of our UNU research team.

As at the La Paz regional meeting, UNESCO had requested the member countries to provide, prior to the meeting, official country statements. Herein lay a little-disguised bombshell that almost destroyed the entire meeting. The Afghanistan official country statement carried a frontispiece map of its national borders that included a large slice of north-western Pakistan!

The first two days at Kathmandu were taken over by so many courtesy speeches and

Fig. 8. Traditional houses and irrigated terraces: my first introduction to the supposed evils of deforestation and too many terraces. The beauty of this scene has been totally destroyed subsequently by a real, if unavoidable, evil – the spread of unplanned urban blight (1975).

Map 1. Outline map of part of south-central Asia showing the countries involved in the UNESCO MAB-6 conference in Kathmandu (1975), together with neighbouring countries.

proforma presentations that peace prevailed. However, one of the Indian representatives had anonymously leaked to *The Times of India* the apparent Afghan territorial claim. This was immediately published with a statement to the effect that India accepted the report. Chaos ensued. The Pakistani ambassador to Nepal lodged a stern protest and demanded an apology. As chair of the meeting, Ratna Rana was placed in a very difficult position.

The Pakistani delegation walked out of the meeting, leaving the Nepalese acutely embarrassed. In deep distress, Francesco di Castri worked with Ratna Rana to organize a temporary adjournment. Fortunately, small working groups already had been set up. One was to prepare a draft recommendation and

functional outline for establishment of an institution to undertake responsibility for an integrated approach to regional mountain development. Kamal Shrestha was appointed as its chair, and there were members from each of the regional countries present. I was asked to serve as reporting secretary. Over the next two days, while Francesco was tied to the telephone link to Paris,[32] we proceeded with the task of writing the first draft of a document outlining tentative functions that eventually led to the formation of the International Centre for Integrated Mountain Development (ICIMOD).

Despite the adjournment of the main meeting, tension remained high. During a reception at the United States embassy the same evening, several Indian participants hassled the two Pakistani delegates who, with no designated place to pray, had laid out their prayer mats in a relatively quiet corner of the large reception room. They were quickly led into a small adjoining room.

Cultural differences continued, even at the personal level. My new colleague Kamal, as working group chair and with me as secretary, determined that I should write whatever he dictated. He assumed he could include in our report a specific recommendation that Kathmandu should be the location for the mooted international mountain centre. This was my first, and by no means my last argument with Kamal, although at the time I had no idea that we would be working closely together for several years. I eventually convinced him, but only after a heated debate, that such a specific recommendation was unthinkable under the politically charged circumstances. The Indian participants were already struggling amongst themselves for the designation of Simla, Dehradun, Nainital, and Darjeeling, all Himalayan centres of learning and prestigious remnants of British India (the Imperial Raj). And, of course, they were united in their opposition to locations such as Gilgit or Islamabad in Pakistan. There was even an Afghan murmur in favour of Kabul.

After considerable effort, I convinced Kamal that, given the furore surrounding the entire meeting, it was our responsibility to write a neutral report and hope that the meeting could be reconvened rather than scrapped entirely. I had argued that lack of a recommendation for a specific location would automatically produce triumph for Kathmandu because political uproar would follow any suggestion of an Indian or Pakistani site.

Following a great deal of politicking, the full meeting was able to reconvene, but there remained only one day to formulate some unanimously supported resolutions (unanimity was the UNESCO way). Yet an invitation from the Government of Nepal for Kathmandu to host the proposed international mountain centre was printed in the final report, although it had not been announced publicly during the final plenary session. If it had, the meeting most certainly would have collapsed (UNESCO-MAB 1977).

All the participants reassembled for the closing session, anxious that there would be time to discuss and approve the various resolutions that had emerged from the in-camera efforts of the working groups. The sense of relief that we had apparently overcome the diplomatic confrontations permeated the assembly. However, peace had not been achieved. We had barely begun consideration of the working group reports when a final, and possibly lethal, interruption occurred.

The door flew open and an imperious figure marched into the centre of the room (delegates were seated in a large circle with observers making a second, outer, circle). The intruder, the Pakistani ambassador to Nepal, addressed Dr Ratna Rana in a threatening manner, demanding both explanation and apology for the fracas caused by the anti-Pakistan report in *The Times of India* (despite the fact that Dr Rana had had nothing to do with the Indian assault). Paul Baker, seated next to me behind the small desk-size United States flag, muttered in my ear, "Now we will likely lose all our vital agreements." Instinctively and involuntarily, I asked Dr Rana, as chairman, to identify the "stranger" who had entered the closed room without invitation. He showed acute embarrassment, but before he could utter a word, the ambassador announced that he had been inexcusably insulted by being referred to as a stranger, noted my nameplate, and loudly declared that he would report my conduct to the U.S. ambassador to the UN. Thereupon he turned and stormed out of the room.

I was taken aback, although there were audible sighs of relief. Our chairman, without comment, quickly picked up on the business in hand. All resolutions were passed unanimously and the meeting was adjourned. During the following lunch break Paul said that he was sure that my "insult" had not been intended but it certainly had been remarkably effective. He thought that any formal complaint to U.S. Ambassador Moynihan would earn me a note of congratulations as he had been particularly intemperate over the widespread petty complaints emanating from UN member delegations, especially ones aimed at the United States. I heard nothing further.

The Kathmandu MAB-6 regional meeting produced few immediate results. Nevertheless, a large gathering, representing an important array of Himalayan countries, UN agencies, and INGOs had endorsed the concept of creating an international mountain institution. Full fruition followed a tortuous route and led to the inauguration of the International Centre for Integrated Mountain Development (ICIMOD) in Kathmandu in 1983. The draft outline that Kamal Shrestha and I had submitted eventually emerged as ICIMOD's founding charter after it had received numerous corrections, deletions, and additions during its meander through UNESCO and German, Swiss, and Nepalese channels. Nevertheless, a critically important mountain institute would arise from the shambles of that Kathmandu MAB-6 meeting.

Building Awareness of Mountain Cultures and Relationships

Although it appeared at the time that there were no immediate results arising from the Kathmandu meeting, a number of very worthwhile initiatives were proposed. As occurred after the La Paz MAB-6 meeting, there were strongly endorsed pleas for compilation of a regional state-of-knowledge publication and for the need for multidisciplinary and interdisciplinary approaches, for interdisciplinary applied research, and for mountain system education and training at all levels. There was also a plea for collaboration between the human and natural sciences. Perhaps the main message was the need to propagate awareness of the rapidly growing threat of serious environmental damage and, further, that such damage in the mountains would likely have disastrous impacts on the subjacent lowlands – the life-support systems for hundreds of millions of people (that is, the lowlands of the Indus, Ganges, and Brahmaputra rivers). Thus, once again, the rising crescendo of imminent Himalayan catastrophe was introduced, this time as a major focus. This "catastrophe" scenario would continue

Fig. 9.   The picturesque setting of the water bearer disguises the warning: "Don't drink the water, this is Kathmandu".

as a theme in the following years. Meanwhile, the UNESCO staff members and consultants from outside the region together urged serious efforts to alleviate mountain poverty and to help ensure preservation of the rich mountain cultures.

The emphasis on the hazards resulting from highland–lowland interactions[33] reflected the concerns expressed during the 1974 Munich workshop. These anxieties were deepened in Erik Eckholm's subsequent publications in 1975 and 1976. Thus the growing

alarm that emerged in Kathmandu quickly led to what I later characterized as the theory of Himalayan environmental degradation. The elements that became flashpoints for Erik are clearly evident in the final MAB-6 Kathmandu report (UNESCO-MAB 1977):

> *Improper use of these mountain lands has led to extremely acute soil erosion and consequently to the danger of decreasing productivity in the whole area, at a time when the rate of population increase is unusually high. Downstream,*

in particular in the lowland plains, the reduced water and soil capacity of the mountains causes great damage by exacerbating annual floods and increasing the rates of sedimentation, especially during and immediately after the monsoon season. The drying up of rivers used for irrigation in the dry season is yet another possible consequence. In sum, human pressures on the high mountain ecosystems are increasing everywhere and threaten the life-support capacity of the

mountain lands as well as the productivity of the surrounding lowlands. (p. 15)

In some areas, firewood has become so rare that animal dung is being increasingly burned for fuel. This creates another problem of reduced productivity of farmland in these areas where animal manure often represents the most widely-used fertilizer. (p. 29)

Some experts have observed that this process has evolved to the extent that the topsoil being washed down into India

Fig. 10. Red peppers are drying in a village square close to Kathmandu.

and Bangladesh can ... be considered as Nepal's most precious export, but in a doubly destructive manner, since it does no good to India and Bangladesh, only harm from increased siltation. (p. 29)

These and similar statements, made the more sensational by Eckholm's colourful and persuasive writing, especially when spread worldwide by the news media, could be regarded as the flashpoint for 20 years of over-dramatization, confusion, and misdirection of policy. In contrast, after more than 30 years of personal retrospection, I am relieved to recollect that the short text I submitted in Kathmandu as a possible annex was included in the final report as Annex 7. I referred to the impacts of the 1968 catastrophic floods in the Darjeeling Himalaya, after which the meteorological officer confirmed that according to his records such an event had occurred only twice previously since records began in the late nineteenth century:

> This may mean that we are dealing with a catastrophic event that has a recurrence interval of between 100 and 500 years. In practical terms this would also mean that the extensive deforestation which characterizes this area will certainly have heightened the magnitude of the event, but we do not know what the difference would have been in the amount of debris flows and mudflows if the mountain slopes had been protected by virgin forests. (p. 74)

The annex also contains my cautionary note on the need for reliable data:

> ... the question of 'how much?' must be answered if intelligent long-term responses are to be developed. Thus, it is not sufficient to argue that soil losses in the mountains, with their attendant and perhaps more serious downstream effects of flooding and siltation on the Indo-Gangetic Plain, are enormously higher, for instance, than prior to the population increase. It is necessary to know by how much. Without a quantitative answer short-term policies, themselves of a very expensive nature, could be put into effect without any sure prospects of being viable in the long term (p. 74).

The beginnings of this early growth in my concern for the tendency to over-dramatize must be balanced with the realization that the entire Himalayan region certainly was facing a formidable degree of mismanagement, exacerbated by the nature of its climate as well as the inevitable political friction. The landslide (debris flow) danger was forcefully impressed on our minds during an early morning excursion to witness sunrise on the Himalaya from a nearby mountain viewpoint. Our bus was trapped by a large rockslide, and many of us joined the local farmers who came to our rescue to help roll boulders off the road in the rain while debris continued to pass amongst us. Our Nepali hosts were horrified to see their Western colleagues rolling up their sleeves and catching a soaking along with the lower caste peasant workers.

Incidents of this kind occurred on several occasions during the following years, reinforcing our awareness of the great differences in work ethic that separated many Western participants from our Asian counterparts. I and my colleagues and students would instinctively roll up our sleeves and get our hands dirty. In the Indian subcontinent (and China proved no exception at the time), our new colleagues were conditioned by centuries of tradition. Of course, the caste system was still very much alive. The situation required a degree of care and the need to develop a strong sensitivity.[34]

I shared another minor adventure with Paul Baker. This was a cautionary experience in terms of choice of a Kathmandu taxi. On a free afternoon, Paul suggested that we should seize the chance and drive to the Tibetan (Chinese) frontier. Accordingly, I had my first experiences of passing through and taking tea in tiny picturesque villages in the Middle Hills above Kathmandu. We reached the frontier and photographed the Friendship Bridge and its People's Liberation Army guardians. On our return we had three separate flat tires and only a single spare wheel. From this I learned the Nepali emergency response of packing dry grass into the deflated tires. We eventually reached our hotel well after midnight.

The chance to visit Kathmandu and to take part in a multinational conference, unsuspected by me at the time, proved something of a crucial awakening to the politics involved and to the need to negotiate amongst a host of different cultures and castes. Despite all the problems and political landmines, the Kathmandu MAB-6 meeting was a revelation of the extremes in the ways of life. The smiling faces amidst abject poverty in the outlying villages, the beautiful and well-maintained terraces, the supposed need for foreign "experts," together with thoughts going back to the 1968 Darjeeling rainstorm, left me puzzled. And Kathmandu in 1975, still with barely any "development" was a most attractive place to visit. Walking through the narrow streets around the Durbar Square at night, illuminated only by torches and charcoal braziers and with little motor traffic, was fascinating.

Fig. 11.   Kangchenjunga, the world's third highest mountain as seen
from Darjeeling. My first view of the Himalaya during the International
Geographical Congress held in New Delhi (1968).

CHAPTER 4

Darjeeling: Personal Awakening to High Mountains and Refugees

My visits to Darjeeling between 1968 and 2000 had such a profound effect on me personally, influencing at least two of the major themes in this book, that I have chosen to bring an account of the visits together in one chapter, although I only detail the more significant of the several visits. It will become obvious that they relate to most of the other chapters. They include my first impressions of the Himalaya in 1968, some years prior to any significant development of tourism and trekking. This early visit occurred shortly after a torrential rainstorm that caused thousands of landslides, much loss of life, and extensive property damage. The devastation stirred the geomorphologist in me, but also my inborn scepticism, and I first began to question the credibility of the claim that extensive deforestation was the primary culprit.

This scepticism of conventional wisdom is a recurring theme of personal awakening. Thus, when Professor Walther Manshard, newly appointed UNU vice-rector in 1976, asked me to take on the task of coordinating one of his planned new projects, I agreed on condition that it include mountain hazards mapping in the Himalaya. As it turned out,

the ensuing research was undertaken, not in the Darjeeling region of India, but in Nepal because of unrestricted access to the high mountains (Chapters 7 and 10).

The second theme is a much more personal one. It derived from a series of coincidences, especially from a chance association with Darjeeling's Tibetan refugee community. This introduction to the education of young refugee girls brought me to a fuller understanding of the circumstances of minority mountain peoples, their customs, close association with nature, and the severe difficulties they faced in trying to assimilate to their new surroundings. It also led me to believe that small personal "aid" projects that require insignificant funds may have a greater chance of success than some of the multi-million-dollar institutional undertakings. The latter are often driven by politics, by institutional competition, and by lack of sound planning. Incorrectly conceived "development" policies can also result in a great waste of funds and even harm the intended beneficiaries.

In practice, the two themes became mutually reinforcing. The understanding

J. D. Ives, *Sustainable Mountain Development*, https://doi.org/10.1007/978-3-030-96029-2_4

emphasizes the present-day widely held view that mountain people must be treated as intelligent equals and play a primary role in decision making.

First Views of the Himalaya

My route to Darjeeling in November 1968 was by a PanAm flight from Denver, Colorado, across the Pacific with stopovers in Hawaii (to photograph volcanic landforms for the Canadian National Film Board), Japan (Tokyo for Mt Fujiyama and Sapporo, as guest of the Hokkaido University's Institute of Low Temperature Science), Hong Kong, and Bangkok. A Thai International flight to Calcutta (Kolkata) with an early morning arrival was timed to connect with an Air India flight to Bagdogra, the local airport for Siliguri and Darjeeling, the latter some 2000 m (7,000 ft) up the Himalayan front range.

The University of West Bengal was official host to the IGU geomorphology commission symposium that was part of the International Geographical Congress to be held in New Delhi. Twenty-five participants were to be housed on the main campus in Siliguri for the first two nights, with the intervening day devoted to an excursion along the base of the foothills. A lecture by Professor Dresch, our commission chairman, was arranged for the evening. Then we were to proceed to Darjeeling for the main paper presentations and field excursions.

Two weeks before I left Colorado, the news media had detailed a exceptional rainstorm with precipitation ranging from 60 cm to over 120 cm on the Darjeeling Himalayan front range (the Darjeeling hills) during a three-day period at the end of the monsoon season. It apparently had caused thousands of landslides and considerable loss of life. On arrival in Siliguri, our group of enthusiastic geomorphologists were very apprehensive whether we would reach Darjeeling. In fact, our welcoming Indian hosts informed us that the original plans had to be changed and the main paper sessions would be held on the Siliguri campus. Then, if the narrow twisting road up the 2000-metre climb were cleared of landslide debris in time, there would be a short visit to Darjeeling later in the proceedings.

With my colleagues Fritz Müller and Barry Bishop, I came to doubt that information. It only took one disastrous day along the hot and steamy terai lowlands, disrupted by the breakdown of half our motorized support vehicles, to increase our sense of need to reach higher altitudes. Furthermore, rumours from townspeople we had met incidentally indicated that the road to Darjeeling was passable and that access was more dependent on the attitude of the Indian military than on the landslide problem. The People's Liberation Army was encamped on the Jelep-la Pass above and within sight of Darjeeling, and nervousness was still high in the aftermath of the Sino-Indian border war of 1962.

Fritz, Barry, and I were convinced that we could charm our way past the Indian military checkpoints although we could not persuade our hosts to make the attempt. Somewhat unceremoniously, therefore, we abandoned our group, rented a taxi with a driver who was

more than willing to attempt the journey for double the regular fare, and set off full of hope.

The drive to Darjeeling required a long and exciting day. The 70-kilometre road with its impressive hairpin bends, had been assailed by more than 90 landslides, some very extensive. Soldiers were everywhere, always courteous, but occasionally requiring a great deal of cajolery to allow us to pass. Barry proved to be most effective at persuasion. We coaxed our taxi across some formidable ground, surrounded by hundreds of peasant workers trying to fill in the road cuts and stabilize the slopes above. We were eventually ushered into the government guest house, the only group to reach Darjeeling that day.

We spent the following day retracing our route for several kilometres on foot to inspect one of the colossal landslides. Hundreds of sari-clad girls and young women were hauling on their backs loads of gravel and rocks from far down the slopes below us to supply their menfolk with material to fill the road breaks and stabilize the slopes above. They sang cheerfully in time to their heavily laden uphill struggle, then skipped their way down for yet

Fig. 12 [top]. In October, 1968, more than 100 cm of rain fell on parts of the Darjeeling Himalaya in three days. The road from Siliguri to Darjeeling was cut by landslides in 92 places. "Thousands" of landslides was the best rough estimate that could be attempted. To what extent was the damage due to deforestation?

Fig. 13 [bottom]. Where the flood waters burst out onto the plains below (the terai) whole villages were washed away (1968).

another load. Some of them made quite lewd suggestions as they skipped their way past us, adding to the general merriment. Fritz and Barry were veterans of this kind of experience. I was simply bowled over by all the new sensations. I was also conscious that Kangchenjunga, the world's third highest mountain, was hiding behind the clouds, waiting to reveal itself as the afternoon slipped away into the early evening.

Returning to the guest house, we were pleased to find that the rest of the party had arrived, no doubt encouraged by our buccaneering approach of the day before. The local notables had planned a reception for us in the grandiose Victorian-era Gymkhana Club and were relieved that the entire group had arrived.

Early the next afternoon, we all assembled in the courtyard of the Gymkhana Club, most in dark suits and ties. We were greeted by an impressive receiving line; behind them were a group of young women in some of the most colourful saris that I had ever seen. So, one by one, we passed along the line, shaking hands with the mayor, chief military officer, president of the Darjeeling Tea Planters' Association, vice-chancellor of West Bengal University, the Anglican vicar, and finally Mother Superior Damien O'Donahoe, the principal of Loreto College, who had offered to host our commission by providing rooms in the college. Mother Damien, in turn, passed each visitor on to one of the line of young ladies – members of her senior classes, who were to serve as hostesses for the entertainment that was to follow in the building, shining as if the Queen-Empress herself were about to inspect it.

After the multiple handshakes, I was introduced to the next girl in line. Muna Lama explained that she was from Bhutan. It was intriguing for me to meet someone from Bhutan, that mysterious Himalayan Buddhist kingdom, totally closed to the outside world at that time, and which I would never visit but with which I would develop a powerful link in later years.

Muna tentatively led me inside and we joined Barry and Fritz, along with Leszek Starkel, whom I had met in arctic Sweden in 1960, also accompanied by colourful hostesses. In the grand ballroom, a small orchestra on a raised platform was quietly playing tunes reminiscent of the Raj and long ago. There was a well-stocked bar, contributed by the local tea planters and offering everything from gin to lemonade and manned by smart attendants. Mother Superior and two of her senior teachers formed a quiet guardian line some fifteen paces from the bar, eagle-eyed, watching every move.

Hesitatingly we made small talk with the girls – Where were they from? What were their favourite school subjects? What were their plans after graduation? and so on. Muna's family were large landowners in Bhutan. She spoke flawless English and when I asked what was her mother tongue, she was slightly shocked – quite English, of course! Other girls were from Bhutan, Sikkim, various parts of the Indian Himalaya, and Calcutta, all apparently from well-to-do families.

Barry, while eying the bar rather conspicuously, asked, "What are we waiting for?" So I responded by asking Muna if I could bring her something to drink. Her reply: "No thank-you, we are not allowed to drink, but I must bring something for you." I had not expected that she would ask for whisky, but surely a lemonade would have been in order? She explained that the girls had received strict instructions before leaving the college, one being that they must have nothing to drink, and nodded towards the three "guards" standing alert. This prompted a mischievous thought. We had a ballroom, an orchestra, and enough couples. Why not dance? The suggestion was met by a mixture of shock, smiles, and giggles, and a confession: "Dancing is part of our deportment lessons in the college, but we have never danced with gentlemen."

A plan quickly evolved based on safety in numbers. We would ask the leader of the orchestra to play a Strauss waltz (the limit of my dancing abilities). Then, when the music began, we would all strike out together so that no individual girl could be identified as the instigator and, if necessary, I would confess and apologise to Mother Damien. It worked beautifully and eventually nearly all the geomorphologists and their hostesses joined in. After the first frowns of disapproval (or was it alarm?), even Mother Damien smiled and swayed slightly in step with Johann Strauss. She was human after all, and I was to learn over the following years how very human she was.

By eight o'clock, Mother Damien and her aides rounded up the girls to be marched crocodile fashion back to the college and, no doubt, with the gates locked behind them. Fritz, Barry, Leszek, and I walked a long way round to the guest house, now under an almost cloudless night sky shot through with diamonds, Kangchenjunga a faint white presence commanding the northern horizon. What an unreal beginning for a symposium on mountain geomorphology!

There followed three half-days of paper sessions with much appreciated tea and coffee breaks in the teachers' common room. During the afternoons, there were a number of local excursions including one to the Tibetan Refugee Self-Help Centre that Fritz urged me not to miss, so we went together. Darjeeling was refuge for many of the Tibetans who had fled their homeland at the same time, or within a few years of the flight of the Dalai Lama.

The self-help centre was designed to enable some of the more skilful artisans to help earn their living. I was so impressed with the friendly reception and the artistry and great range of the work under way that I sought out the manager and proffered a modest financial contribution. With utmost courtesy, he thanked me and explained that to accept would be against the principles of the centre, but if I wished to purchase, for instance, a rug …? So I purchased two rugs and found they could be shipped to Colorado for the very modest price of US$64 each, including shipping.[35]

The manager alerted us that we might be approached on the street by Tibetans who were seeking direct assistance from tourists and the "come-on" was a request for cash to help with education of small children, and that we should be careful. He knew of several over-anxious mothers who had persuaded multiple tourists to give money for the same child. This led me to ask Muna if she could

Fig. 14.   The Darjeeling Tibetan Self-Help Centre produced outstanding hand-woven rugs amongst a great variety of traditional ware. "Please do not offer charity, but do purchase a rug – or two" (1968).

Fig. 15.   The main congress of the International Geographical Union was held in New Delhi and opportunities for local excursions were plentiful (1968).

help by locating a needy family whom we might assist.

Another excursion for me and Fritz was to visit the local meteorological station. One of the overwhelming mountain geomorphological impressions was that the torrential rainstorm had caused thousands of landslides. It was impossible to count them. We certainly took many photographs and I hazarded a guess that there were more than 20,000 slides in the wider Darjeeling district (Ives 1970).

Our Indian leaders had insisted that the cause of the devastation was the extensive local deforestation that had exposed the steep slopes to the full impact of the rainstorm. From the beginning, I doubted this facile explanation, the more so when the officer in charge of the meteorological station

responded with confidence to a series of questions: the rainstorm of the recent past, he informed us, was only the second event of such magnitude since the station was set up by the British almost a hundred years ago. The earlier rainstorms had caused innumerable landslides despite there being far more forest cover than today. This was my first, if accidental, introduction to this form of catastrophism that almost came to dominate later thinking worldwide. Implanted in my mind this way were seeds of controversy that influenced my thinking for decades afterwards.

Finally, before saying goodbye to Mother

Damien and Muna, I sought a private interview with the Mother Superior and was invited to take tea with her. As a non-tea- drinker, this was one of the small number of exceptions (China, much later, was an instance of even greater submission). The setting for the meeting was Mother Damien's study – warm wood panelling, basketwork armchairs, and mullioned windows through which the form of the majestic mountain could be seen. Muna and a colleague brought in the tea paraphernalia with hot scones, butter, and strawberry jam.

Immediately after the girls had left us, Mother Damien surprised me, starting the conversation by expressing her pleasure that the girls had had the opportunity to dance with us at the Gymkhana Club. She then gently made it apparent that she was expecting a request from me. It was simple. I asked that, if I were able to locate a small Tibetan refugee girl, would it be appropriate for her to receive an education at the college. I was warmly encouraged and advised that I could depend on Muna to assist in the selection.

The conversation then turned to everyday matters although I could not resist telling her that my mother's family were originally from Cork, Ireland. Without a moment's hesitation, she smiled and remarked that my colouring matched the location but she suspected we were not members of her Church. She also informed me that she thought Muna was highly talented and hoped that she could eventually attend university: Would I please assist if there were any way I could? From this brief interlude a strong friendship developed that lasted until after her retirement and return home to Ireland many years later.

The Darjeeling symposium had several very marked effects on my future. Through Muna I found a very worthy Tibetan refugee family, to be described later in this chapter. I left Darjeeling with a controversial notion about deforestation and the impact of exceptionally heavy rainstorms. This became

Fig. 16.   December, 1968, in the Hindu Kush. The proprietor of a bazaar in the village, upon hearing my accent, insisted that, as an Englishman, I must purchase a beautiful curved sword guaranteed to have cut off the heads of British officers on the Khyber Pass. There was a commotion as I entered the USSR, Tashkent, with sword in hand.

an important element of the United Nations University mountain research project that was to be initiated ten years later (Chapters 7 and 10). Regrettably, I was not able to assist Muna, nor could I have known that the problems to be faced by her family, despite considerable wealth, would lead to confiscation of their property and exile because they were Lhotsampas, members of a Bhutanese Hindu ethnic minority that was predominantly of Nepali descent. I did not learn of this problem that was to convulse Bhutan until the 1990s after D.N.S. Dhakal, who also came from Bhutan, joined me in Colorado as a UNU research fellow.

Our days in Darjeeling eventually came to a close. There were warm goodbyes to Loreto College staff and students. I had asked Muna to look for a Tibetan family, preferably one with small girls, as even at that time I had become convinced that there were advantages in assisting with the education of girls. She, in turn, confirmed Mother Damien's comments that she hoped to obtain a university education after graduation from the college.

From Darjeeling we all proceeded to New Delhi and the main International Geographical Congress (Chapter 2). After the Delhi meetings, I continued on the round-the-world tour set up for me by Dean Jim Archer as part of his efforts to persuade me to leave Canada and take the position with the University of Colorado. This final section of the tour was decidedly unconventional for the time: Kabul and the Afghanistan Hindu Kush mountains, Tashkent and Samarkand, Moscow, London, a visit to my parents in Grimsby, and then home to Boulder, Colorado.

Return to Darjeeling in 1975: The Tibetan Refugee Connection

The newly acquired Tibetan handwoven rugs reached Boulder a few weeks after my return from Darjeeling. It quickly became apparent that I had a small local market demand on my hands, and several neighbours were prepared to pay double the set price I had paid. A brisk business developed, and the "added value" was transferred to the Tibetan Refugee Self-Help Centre to assist with the education of their children. I managed to sell over 50 rugs.

During the next several years, letters were exchanged with Muna. She located a Tibetan family whom she strongly recommended. By this time, I had become involved with the UNESCO MAB-6 project and, as a regional meeting was scheduled for Kathmandu (Chapter 3), I was able to visit Darjeeling again and to meet with the Tibetan refugee family.

The main focus of Muna's letters, however, was her enthusiasm to undertake university-level education and her father's willingness to support such a venture. I managed to make arrangements for her to take a bachelor's degree, majoring in geography, in my own department in Boulder. Then there was a surprising turn of events. Muna was to be married to an Indian cousin. Pauline and I were invited to Bhutan to the wedding but were unable to attend. I have always regretted that we did not attend, and this regret grew with the passage of time when I learned about the expulsion of the Lhotsampa minority from Bhutan. Muna was married and moved to Calcutta (Kolkata) where her husband was employed.

The next step in the Tibetan relationship was abetted by my invitation to represent the United States at UNESCO's regional MAB-6 meeting in Kathmandu in September 1975 (Chapter 3). I had already corresponded with Mother Damien to explain that I was planning to meet the Tibetan family that Muna had identified for me and hoped that I would be able to visit Loreto College. However, en route to Darjeeling I had the opportunity for a two-night stopover in Calcutta during which I visited Muna and her husband and dined with them. After dinner, they took me walking through parts of Calcutta that I would never have included on my itinerary. They felt I should witness one of the desperately poor sections of the city before I entered the luxury of my hotel. I was suitably shocked. One image that will remain with me as long as I live – deliberately mutilated small children chained to broken pavement blocks as captive beggars.

The next morning, I took a flight to Bagdogra and an exhilarating taxi ride up the renowned road to Darjeeling and to the Everest View Hotel. The following morning, Norbu, the eldest son of Tshering Choni, father of the family Muna had located for me, came to the hotel. He guided me to meet his parents and siblings and share a midday meal with them. We walked together, somewhat awkwardly, across the town to a ridge overlooking a pronounced valley with Kangchenjunga rising above, although at that time of day, it was lost in a heavy shroud of cumulus clouds. Along the ridge was a line of tiny houses – perhaps

one should say "huts." We approached one near the middle of the line to find a traditionally dressed Tibetan family clustered around the doorstep, anxiously awaiting my arrival. Norbu, who spoke some English, formally introduced me, first to his father, Tshering Choni. I was ushered into the main room of their small house and was seated on a low dais. The walls were hung with *thankas* and various Tibetan decorations. For me, it was like entering another world. A nephew of Choni's had been invited as interpreter.

Then began one of the more exceptional experiences of my life. Choni had been a farmer-tailor in a small village not far from Lhasa. He explained, through his nephew, that because he had a comfortable living, when the Dalai Lama fled from Tibet he decided to stay on. But within a year, individual males in the village began disappearing during the night, presumably taken by Chinese soldiers. When his own brother was taken, Choni decided that it was time to leave. By that time, the People's Liberation Army had sealed all the main passes through the Himalaya, so Choni and his wife had to cross by a high pass with the two small boys, Norbu and Urgen, on their backs, together with very few of their possessions. En route to Darjeeling both parents had frozen their feet, although they later recovered.

I learned the names of the two younger daughters, Lhazom and tiny Yudon (nicknamed Yula), who were shy and smiling, hiding behind their mother's skirt. The girls had been born in Darjeeling. We exchanged information about our families. In the process, I asked the birthdates of the children and was astounded as I wrote them down that all seemed to have been born on the 15th day of different months. Perhaps, I thought, this was something to do with a Tibetan birthdate-naming tradition, although I was aware that the Tibetan calendar was different to my own. My mother and father had birthdays on 15th and so did I, and so "15" has always been a special number in the Ives family. I was over-whelmed when told that Yula was born on 15 May, the same day as Peter Ives, our youngest son. I had not mentioned my family connec-tion to the number 15 until after I had been given their birthdates.

Then followed a really intriguing ques-tion. Choni asked whether I had at least a faint recollection that we had met two reincarnations ago. I had to confess that my memory did not stretch that far back. He then explained that we had indeed met and we each had agreed to take care of the other's family if and when the need arose.

A meal followed, including abundant traditional tea (sour butter and salt) with which I struggled valiantly. Finally, it was time for me to leave. Choni took off the wall one of the only two *thankas*, precious family heir-looms that they had carried out of Tibet. He presented it to me with much ceremony, say-ing that it was very old and should be mount-ed above our "marriage bed" as it would ensure fulfilment. Accompanied by Norbu partway, I walked across the town in a daze, barely waking up in time for tea with Mother Damien at Loreto College.

Mother Damien greeted me like an old friend. I was seated in the same armchair with the fractured view of Kangchenjunga emer-ging from the clouds. We talked briefly about Muna. Her first question was whether Muna had a piano in her Calcutta home. I had to say no. She then expressed her sadness and told me that, while a student, Muna had shown promise of becoming a talented classical pianist.

Then to business. I had decided that, with Mother Damien's approval, Yuden would at-tend school at the college, first in the elemen-tary section and then through high school if

Fig. 17.   Tshering Choni and family, with the exception of Urgen who was in Kathmandu attending lessons with a Buddhist monk master painter. The children are, from the left, Norbu, Lhazom, and Yuden (1975).

Fig. 18. Urgen soon became a proficient thanka painter. Here he is working at home on one of his early masterpieces.

Fig. 19. An example of Tshering Urgen's finished work. The mountings were sewn by his father who was an expert tailor/farmer before the family fled from Tibet.

her progress justified it. But there was a rider attached to the agreement: Could I accept the extra responsibility of supporting Lhazom (Mother Damien was obviously familiar with the Tshering family) as it would be hard for the older sister if little Yula received all the benefit of my visit? I agreed, feeling I should have thought of that myself. However, I also had a "condition": that they be day-students and live at home. Mother Damien initially disagreed. I said there were two related points: first, I did not want to risk separating them from their Tibetan family; second, I did not want any possibility of conversion to Roman Catholicism. She laughed at that and explained that she was aware of my Unitarian proclivities but insisted that the college policy was never to seek conversions: "We demonstrate our views to our students by example alone." To which I risked offence by commenting, "That was what I feared!" She quietly laughed, poured more tea, and said she thought we would get along very well in this venture.

The last part of the arrangements were completed a couple of days later in Kathmandu. I recounted my experience to Corneille Jest, a ranking Tibetologist. He urged that I must ensure that all the family benefited from the arrangements, not only the girls. For this, I must make sure that they received Tibetan New Year presents.

Tibetan Refugees and Loreto College

The disturbing manner of my 1979 journey to Darjeeling overland from Kathmandu is described in detail in Chapter 7. Nevertheless, after the rigours of the inappropriate choice of route and a restful night in the guest section of the Gymkhana Club, a pleasant day lay ahead. After a late breakfast, Norbu and Urgen, Choni's sons, arrived to take me to their home for a meal with the family. Yuden and her sister Lhazom were in school and I would see them later. By this visit, Norbu's English was very good so that we did not need an interpreter. We had an enjoyable reunion and I endured more Tibetan tea, one of the rare traditional menu items that I find hard to swallow.

In the afternoon, I arrived in suit and tie for my teatime appointment with Mother Damien. She sat me down in her elegant study, and we looked out once more toward Kangchenjunga, which this time was almost entirely lost in great banks of cloud. We discussed Yuden's progress, which Mother Damien thought was excellent; she questioned me about my work with UNU and then called in a teaching nun to give me a tour of the college and visit Yuden's and Lhazom's classroom. The attractiveness of the college always delighted me, and it was a joy to see the girls in their classrooms and to be able to have a short chat with them.

I eventually wandered through the college gates, across the town, and back to the Gymkhana Club. After dinner with Dilip Dey, my colleague from the first MAB-6 meeting in Kathmandu, and several of his friends, we had a long discussion about problems facing the Himalaya. The next morning, Dilip drove me to Bagdogra airport and I was soon on my way home via Calcutta, New Delhi, and Frankfurt.

The course of Yuden's and Lhazom's

education was a long one and I made several subsequent visits to Darjeeling. On one particularly memorable occasion after I had visited the class rooms and talked with the girls, I sat down with Mother Damien for tea. After opening greetings, she looked at me quizzically for what seemed a long moment and then spoke very softly, "Yuden is a very clever girl. She wants to be a doctor. I am afraid the cost of her education is going to far exceed your original expectations but I do hope you can provide the necessary support." How could I refuse? She explained that Yuden would need to take biology classes, and Loreto College had none to offer. As a first step, she would have to transfer to the Australian Grammar School and prepare for Cambridge Higher School Certificate. The Australian school was twelve miles away, so it would require full residence.

That was the beginning. Eventually in 1983, with the advice of an influential mountain friend, Joe Stein, an American architect with business headquarters in New Delhi whom I had met at the 1974 Munich conference, Yuden entered one of India's best medical schools, Lady Hardinge Medical College. She eventually obtained her MD and went on to specialize in gynaecology, completing her internship in northeast India. She returned to Darjeeling to devote her skills to working in the poverty-stricken areas of Darjeeling.

In the meantime, Urgen had become a talented *thanka* painter, and I was able to sell quite a number of his *thankas* to students and friends in Boulder. The proceeds assisted Choni in establishing, first, a small restaurant, then a small hotel, which grew into a highly respectable enterprise. Norbu and Lhazom entered the business, and the Tshering family flourished.

In 2001, on the occasion of another mountain conference in Kathmandu, I was able to visit the Lhotsampa refugee camps in eastern Nepal. Dr D.N.S. Dhakal, my former UNU graduate student in Colorado who had become a refugee camp leader, proposed that he accompany me to Darjeeling. This time we would be able to stay at the Tshering family hotel, which he had visited previously and he was anxious for me to see the family. A highlight of the visit, of course, would be to see Dr Tshering Yuden in her work environment. Regrettably, riots in the Siliguri area forced us to turn back. However, we were able to meet the entire family on the Nepal side of the frontier in the hills above Biratnagar, where we spent an enjoyable day together.

The connection with Dhakal, as he prefers to be called, is also a long story that leads into the expulsion of the Lhotsampas from Bhutan and the Ives family's second involvement with the education of refugee girls – this time, Hindus who, together with their families, were in 2011 admitted as immigrants to the United States. The U.S. government agreed to settle a large proportion of the 120 000 Lhotsampas. Other countries, including Canada, are making provision for the rest.

Fig. 20. International Geographical Congress in Moscow, 1976. View from my hotel onto St Basil's Cathedral and Red Square, which I had suggested as a possible surreptitious meeting point with Mischa Plam.

CHAPTER 5

Moscow and the Caucasus
The IGU in the Cold War

As one of the incidental and highly personal benefits of getting together with colleagues during the La Paz and Kathmandu MAB-6 conferences in 1974 and 1975, I came to know Paul Baker. He urged that I follow his example of the previous year and apply for a John Simon Guggenheim Memorial Fellowship. I never would have thought of doing this without Paul's prompting.

I took Paul's advice shortly after my return from Kathmandu to Boulder. For referees I approached my old Ottawa colleague, Dr James M. Harrison, former director of the Geological Survey of Canada and currently assistant director-general of UNESCO, and Professor F. Kenneth Hare, who had originally encouraged me to emigrate to Montreal in 1954 and take my doctoral degree at McGill University. The third, naturally, was Paul himself, a recent fellowship recipient. And I chose Bern University, Switzerland, as my year-long host institution following the enthusiastic encouragement of Bruno Messerli, thereby further cementing our IGU and MAB-6 collaboration.[36]

While in the midst of preparing the Guggenheim application, I had received a letter from Moscow. Dr Rimma Zimina, one of my IGU commission colleagues, had written to explain that her husband, Academician Innokentiy Gerasimov, would be leading a team of senior Soviet scientists on a cross-country tour of the United States the following spring. They would stay three days in the Boulder-Denver area. Would I please invite him to be a house guest with my family so he could have some respite from officialdom and the anticipated heavy CIA security?

In early May, I arrived at Denver's Stapleton Airport to meet the Soviet delegation. Gerasimov separated himself from the large throng that emerged and took me aside: "Can you get me away from the crowd as quickly as possible?" So we set off at a brisk walk towards the exit (baggage was to be delivered by an assistant). But we were rapidly overtaken by a vigorous young man in a dark grey suit. He smoothly identified himself as CIA and requested a brief quiet word with me. I excused myself from my guest and bent my ear. He quietly explained that he was aware that I was not an American citizen but hoped that I would cooperate: the message was more or less, "Do not take our visitor more than 15

© The Author(s), under exclusive license to Springer Nature Switzerland AG 2022
J. D. Ives, *Sustainable Mountain Development*, https://doi.org/10.1007/978-3-030-96029-2_5

minutes beyond reach of a telephone, record any contacts he may make or receive, and do not fail to be back with him at the airport well in time for his flight departure." With a grin he concluded, "If he doesn't catch the scheduled Aeroflot flight from JFK, my head will roll."

So Innokentiy stayed in Boulder as a family member for three days. It was an intriguing and most convivial experience. Our four children were much impressed, especially Nadine, our eldest, who at 17 was studying Russian language and literature and – how shall I put it? – was somewhat leftish in her political frame of mind.

There was one especially memorable aspect of the visit. Over family dinner on the second day, Innokentiy paid compliments to our house and the attractiveness of our living room window view onto the Rockies, including the tip of Longs Peak at over 4200 m (14,000 ft). He then turned to me seriously and invited me and the family to join him in Moscow – on a permanent basis! He insisted that our standard of living would be much higher; there would be servants, a limousine and driver, and a much more elevated social standing than we would ever acquire in America. The talk of servants and limousines astonished Nadine. With a red face, she excused herself and precipitously left the room – I am sure that her politics changed substantially on the instant.

After my attempt at a cautious and polite response, Innokentiy assured me that I was misreading the situation in the USSR – no doubt due to American propaganda, but if the entire family were not to come, the invitation would remain open and Nadine would be welcome as a family guest for a year – then her Russian language abilities would be perfected. The final gesture was that I would receive a special welcome when I attended the Moscow IGC later in the year.

After visits to NCAR, NOAA, and my institute, and a personal lunch with Gilbert White, a leading American geographer, I made my way with my important guest to Stapleton Airport with lots of time to spare. This provoked a remark of sincere thanks from the CIA, especially enthusiastic over my list of all our doings of the previous three days. He grinned and said he would see that a "good behaviour" report reached the Canadian security authorities.

The months following Innokentiy's visit proved very hectic. Correspondence was exchanged with Rimma in preparation for the mountain commission's planned excursion to the Caucasus which was to precede the IGC main meetings in Moscow. It was a rare opportunity for Western geographers to visit this region of the Soviet Union, and I was flooded with applications although there were only 20 places available.

There was also my own involvement with the Moscow Congress as a member of the United States delegation and a somewhat alarming request from a friend in Cambridge. Hilda Richardson, secretary-general of the British Glaciological Society, wrote on behalf of the society and a group of senior Cambridge University and Scott Polar Research Institute glaciologists and geographers. They believed that my contact with Academician Gerasimov would provide me with a high degree of privilege, even protection, while I was in the USSR.

In view of this, would I consider acting as their agent in an attempt to obtain clearance for a Russian Jewish glaciologist, his wife, and son to leave the Soviet Union? I sent a nervous affirmative and was told I would receive specific advice later. Dr Mischa Plam had been instrumental in the establishment of the Caucasus (Mt Elbruz) mountain research station that our commission was scheduled to visit. As a Jew, he had come under increasing pressure and had been deprived of any professional employment. Furthermore, his wife, Olga, was the daughter of a former Soviet nuclear scientist, also Jewish, who had been imprisoned by the KGB for many years. It appeared that my visit to Moscow and the Caucasus would have an element of excitement!

My next task was to prepare for a visit to Skaftafell, Iceland. I was keen to take my eldest son, Tony (Anthony Ragnar), with me. He was then 15 and had been named for my colleague, Tony Prosser, who had been lost on the ice cap in 1953, and the farmer, Ragnar Stefansson, whose farm had been the fall-back base camp for our University of Nottingham student expeditions to Vatnajökull, Iceland's largest ice cap. I planned to spend a couple of weeks with Tony at Ragnar's farm and then leave him for a further three weeks while I returned home for final preparations for the year in Switzerland and to make arrangements to leave for Moscow and the Caucasus.

To move the family of six for a full year to Switzerland was no light task, involving housing, schooling, travel arrangements, and shipment of our Volvo across the Atlantic to Southampton. Then Pauline had the heavy task of preparing our Boulder home for rental as well as the intricate arrangements for the children, and for a summer stay based on our parents' homes in two different counties of England. She had to retrieve the Volvo from the Southampton docks, meet Tony's flight from Iceland at Heathrow, and drive with all four children across France to rendezvous with me at the Wild Rose[37] rest stop off the highway south of Bern at 3 p.m. on a pre-arranged day.

Journey to the Soviet Union

After returning from Iceland and assisting Pauline with the final efforts to ensure a smooth family transfer to Appenberg in the Emmental, Switzerland, I departed for Moscow as Pauline and the three children took off for England. Following the Moscow conference, I would travel to Zurich and Bern. There we would take up residence in the beautifully reassembled 200-year-old farmhouse at Appenberg[38] outside Zaziwil. I would undertake some fieldwork based on Bruno's timberline chalet in Canton Wallis and return to Appenberg and from there to rendezvous at that Wild Rose restaurant! Pauline and the children were precisely on time; my coffee was still warm!

While the family was in England, my attention was riveted on the Caucasus field demonstration and the Moscow conference. I arrived in Moscow on the afternoon of 15 July. I found myself somewhat anxious over what was beginning as a long and tedious security procedure amidst milling crowds. I had only just joined the official queue when Rimma burst in, executed the traditional Russian

57

Fig. 21. The western Caucasus on the pre-congress field excursion of the IGU Commission on High-Altitude Geoecology (1976).

Fig. 22. Mount Elbrus, Europe's highest summit, as seen during a subsequent field excursion.

Fig. 23. An impressive 'look-out' along the Georgian Military Highway that cuts through the Caucasus.

Fig. 24. Mount Kasbegi on the southern slope of the Caucasus, famous as the locality where Prometheus was chained as punishment for giving to humans the secret of fire.

greeting and whisked me through immigration officials and Red Army soldiers: "Jack, we have a limousine for you. If you give me the luggage checks I will see that your luggage is delivered to the hotel." Then we began our long drive into central Moscow during which Rimma explained that never again would I be subject to customs inspection. I was taken aback. Should I mention my Cambridge-prompted mission? NO. I assumed that discretion was in order.

So after coffee in the hotel with Rimma, she left me to check in, assuming that everything had been "fixed." However, with Rimma scarcely beyond recall, the hotel reception insisted that I had no reservation. After nearly 15 minutes of haggling I demanded to be shown the reservation list. There was my name, but spelt YVES. Even the top management would not accept my explanation; the spelling was different to that in my passport. It proved necessary to telephone Academician Gerasimov. Within a flash, the management fell over backwards to charm me, even presenting me with a large vodka. I wondered if anyone would spend the next few weeks in northern Siberia after that snafu.

All 20 Western participants showed up for breakfast the next morning. After greetings were exchanged, we took a leisurely meal and departed by bus for Moscow's domestic airport. Here we were greeted by Rimma and introduced to her twenty-some colleagues and various guides, assistants, and secretaries. Rimma presented Olga to me, a young woman who I was told was to be my personal guide, both during the Caucasus excursion and the Moscow Congress. Apparently she

was a friend of the family and Innokentiy thought it would be a good chance for her to improve her English.

Aeroflot flight 1213 TU-154 touched down at Mineralnye Vody on the northern approach to the Caucasus at 11:15 a.m. From there we travelled by bus via Cherkessk to the western Caucasian village of Dombai, a total driving distance of 247 km. En route we made a special stop to visit the Museum of the Defenders of the Caucasus Passes (1943). At this point, my new German colleague, Professor Harald Uhlig, took me aside. Chuckling gently, he gave me a different slant on the heroic World War II exploits of the Red Army and the local people. Harald had served as *Leutnant* in the *Wehrmacht*. He had led his company to one of the high passes. Then all his reserves and most of his weaponry were diverted to Stalingrad. He told me that, had his unit been left fully operational, he expected that he would have been able to cut through to the Caspian oil fields like a knife through butter. But even in the situation where he found himself, he was still able to declare a weekend leave and, with three colleagues, plant a swastika on the summit of Mount Elbruz – "but not a word to anyone else, Jack, please."

We spent three nights at Dumbai and enjoyed two long days of walking in the impressive western end of the Caucasus. On 21 July, we set out on a 400 km bus drive along the northern foothills into the Baksan valley and onto the flanks of Elbruz. We stayed five nights at the Cheget Hotel close to the Moscow State University mountain research station.[39] After lunch on 26 July, we began our long journey back to Moscow, reaching our hotel well after midnight. I am sure all our Western participants were amazed and delighted with what we had seen and how well we had been received.

Special Moscow Treatment for a Canadian Delegate

My special treatment in Moscow included a suite of rooms in the newly opened Rossiya Hotel. They looked out over St Basil's Cathedral onto Red Square and the Kremlin. I was taken there by limousine from the airport. The first day of the Congress (27 July) was occupied with registration and related affairs. Olga, together with an Academy of Sciences limousine and driver, presented herself punctually after a late breakfast and we drove to Moscow State University where the main Congress activities were scheduled.

Once at the university, Olga proved an excellent guide, walking me through the rather extensive registration procedure. She then ushered me into the park where we had a light lunch. We were presently joined by a near-replica of the vigorous young man who had quietly accosted me when I was meeting Academician Gerasimov at Stapleton Airport, Denver earlier in the year. First, a hearty handshake, during which he used my first name as if he had known me for years; then he muttered that he needed to talk with me quietly – could I off-load the girl?

During our Caucasus excursion, I had soon discovered that Olga was an avid consumer of ice cream. So I asked her if she could obtain three large bowls of ice cream, it being

a very warm day. Once she was away, my new colleague explained that as I had missed the prep meeting in Washington, DC for the U.S. delegation, would I be willing to hear one particular point – a warning – despite my not being an American citizen? Of course, I agreed. He went on to say that the Soviets had re-introduced their earlier Cold War tactic of attempting to sexually compromise a certain category of visitor – Third World academics and those from the West who had Third World contacts. I should consider myself a potential target. As an aside, he suggested that I may already be in their mesh with Olga. His advice was as follows: "The morning after, a faceless man in grey will show you a series of compromising photographs. If you can say you think they are marvellous and ask if he can supply extra copies for your friends, then you should go ahead and enjoy yourself. But even if you can say that today, the KGB (the infamous Soviet secret police) has a long memory and you may not be in that happy position ten years from now." As I thanked him and intimated that I thought I was secure on that front, Olga returned with three bowls overflowing with vanilla ice cream. The CIA agent endeavoured to impress upon Olga that we were old geographical colleagues and then he left us.

So we sat in the park and seemingly eyed each other cautiously. Olga broke the silence and asked if I would like to attend a symphony concert at Tchaikovsky Hall that evening. She was well aware of my enthusiasm for classical music and ballet. I asked how could we obtain seats at such short notice? There was no problem. The academy limousine would pick me up at the hotel with time to spare and I should wear my dark suit.

At 7 p.m. as I was waiting in the hotel lobby, a spectacular Olga approached in evening gown and pearls. There was more surprise when we were grandly shepherded to our front row balcony seats despite a fully packed hall. I was quickly lost "On the Steps of Central Asia," followed by Tchaikovsky's violin concerto and Borodin's second symphony. Arriving back at the hotel, Olga also got out of the limousine. My CIA friend's warning crept back into my mind. But all was well – a gracious goodnight and a promise of a visit to the ballet later in the week. To cover the embarrassment brought on by my evil thoughts, I made the gesture of kissing Olga's hand (prompting her to giggle).

Wednesday was the big day – the official opening ceremony in the Kremlin. It was very warm, so Bruno and I had discarded our jackets, assuming that white shirts and ties would suffice. Together with a very large crowd, we had to show our passes to the guards. It was then that I realized that while my pass was red, Bruno's was blue and all those around us that I could see were also blue. I was very slow to understand the significance!

We entered the great hall. I sat with Bruno and the Swiss delegation well towards the rear. To my dismay, amid a group of VIPs sitting on the platform before us was an empty seat. Now I understood the purpose of my red card! Nevertheless, the empty chair was deftly removed, and we sat down to endure two hours of politically correct speeches following the official welcome by Soviet Premier Alexei Kosygin. Academician Gerasimov gave a lengthy and colourful welcome address. Then came a break, to be followed by a period of outstanding entertainment. As we were working our way out of the hall, Rimma came rushing up: "Jack, where were you? You should have been on the platform." I apologised for my unfortunate misunderstanding and explained that I had not brought a jacket with me because of the very warm weather. Even the Gerasimov high influence would not have been able to accommodate shirt sleeves. So to make up, she insisted that after the interlude I take the seat that she had reserved for me on the front row with her and her husband. That I did, but unfortunately having no jacket, I had missed having tea with Mr Kosygin.

As we filed out into the still very warm fresh air, I told Bruno there would be time to walk back to the hotel to retrieve my jacket and suggested that he do the same. And the next time, when we re-entered for the special entertainment, Rimma was waiting. Even then, I was reluctant and indicated that I was with Bruno. So he was invited also. On entering the front row, I noticed that Rimma signalled to her husband. He immediately turned and the person on his left departed to provide space for Bruno. When we all sat down, I found myself between Rimma and Innokentiy, feeling a little ridiculous wearing my field binoculars that I had brought with me anticipating a seat much farther back.

The highlight of the concert was a modern-dress ballet arrangement from Bizet's *Carmen*. The Bolshoi Ballet was spectacular. Into the second act, Rimma nudged me and asked if I would lend my binoculars to her husband "who liked to look at the ballerinas' legs."

Somewhat startled, I obliged. After several minutes of careful observation, Innokentiy returned them, apologising to the effect that I would also want to take a look. I felt obliged to comply and was impressed. We were sitting so close to the ballerinas that the binoculars, focussed on the nearest leg, showed only what appeared as a circle of fine-mesh mosquito netting.

And so the days passed, full of academic and pastime activities, dinners, and short excursions. Olga took me to the ballet, again ravishingly attired. Next we tried the Tretyakov Art Gallery. As she steered me towards the front of a very long queue, I teased her – in a country of free workers, how is it that we do not wait our turn? She had a passion for Delacroix, the great French Revolutionary artist (as I have). Since she had been trying to impress me with how magnificent it was to live in the Soviet Union, with holidays on the Black Sea, concerts in Budapest, dinners in Prague, and so on, I couldn't resist asking her if she had seen the ultimate Delacroix collection in the Louvre. Of course, I was well aware that Paris and the West would have been "out-of-bounds"; she hastily changed the topic.

Eventually Saturday arrived. Hilda Richardson had told me that Dr Plam would contact me by telephone, most likely at the weekend. I excused myself from all activities on the Saturday evening to "rest" in my room. About ten o'clock, the phone rang. I found I was very tense. I picked up the receiver and heard a strongly accented voice claiming that I would know who was calling: "When can I see you?" My tension mounted. I suggested the next morning, ten o'clock, two hundred paces beyond St Basil's in mid-crowd on Red Square (before I had left Boulder, I had been reading Ian Fleming's *From Russia with Love*). He replied that would be too exposed and we would be seen, that it would be best if he came to my hotel room at the same time. This seemed strange to me as the hotel telephone exchange had surely recorded the conversation. Nevertheless, I agreed, but at the cost of a rather restless night.

Shortly after ten o'clock the next morning, there was a knock on my door. I opened to find Dr Plam and a very attractive and obviously Russian lady. A warm handshake from Mischa and an introduction to Olga Plam. Mischa urged us inside. I was very worried. I silently pointed to the light fixtures mutely asking if it were likely that we were bugged. Mischa grinned and expostulated that it was very unlikely, and I should regard myself as "safe" as a special guest of Gerasimov. It seemed that everyone knew of my relationship with Academician Gerasimov – Cambridge, Mischa, the CIA, and Mischa Grosswald, an earlier house guest in Colorado. Nevertheless … We collapsed into comfortable armchairs. From his briefcase, Mischa placed a bundle of papers on a side table. Next he drew out a bottle of Georgian cognac. While he decorked, I produced three glasses. We stood, I very self-consciously, and we made a formal toast to their eventual arrival in Boulder, Colorado.

We sat down again and Mischa began to recite the pattern of his recent history: removal from his academic position as director of the Moscow State University mountain research station, spurious medical examinations, false findings of stomach cancer, forced confinement, and surgery. He showed me the scars on his abdomen. It was like a play-acting of Solzhenitsyn's *Cancer Ward*. I was appalled. How naïve we are in the comfortable West. Did things like this really happen? Yes!

Mischa explained that my mission was to take the documents with me when I left the Soviet Union, send copies to the leading U.S. newspapers and the originals to the White House with the appropriate covering letter noting how I had obtained them. At that period, it was becoming known in some circles that news media announcements of maltreatment would sometimes prompt the Soviet Union to release such Russians "of the wrong colour" (that is, Jews) and to issue exit visas.

We drank another toast. The cognac burned my throat. Warm embraces, and they were gone. Very nervous, I sat and pondered. What if I were to be searched and the documents, obviously defaming the Soviet Union, discovered. I decided on a course of action. In those days, my memory was nearly perfect. I locked myself in the bathroom, read the documents with special care and anxiously tore each piece of paper into small particles and flushed them. Next I took out my passport, airline tickets, and money. I placed a hair from my head carefully across the catch of my suitcase (more notions of James Bond). Then, with only a briefcase and a few personal papers, I fled the hotel.[40]

As I walked the streets in a state of agitation I bumped into Brian Bird, of all people – my McGill University doctoral advisor. He immediately asked me what had happened, saying he could see from my face that something

was amiss. I suggested that we have lunch. Over lunch, I recounted my morning's experience explaining that it had been arranged by our friends in Cambridge. Brian was a St Catharine's College, Cambridge man himself. He couldn't resist pointing out that my limousine and "girl guide" seemed to be missing. Then, with his customary knowing laugh, he said, "Jack, I know your longstanding tendency to tell a good tale, but if you go missing for more than two days, I will report your disappearance to the Canadian embassy. But just think what a story you will have to tell when they let you out of Ljubljanka jail."[41]

After lunch with Brian, I crept back to my hotel sanctuary. Nothing happened. I even had a good night's sleep, was picked up by Olga and limousine the following morning, and proceeded with geographical affairs. Nobody looked at me strangely. I did not learn until many months later that, within ten minutes of leaving me, Mischa had been arrested by the KGB and spent three days in Ljubljanka, the ill-reputed jail and interrogation centre.

On the Monday evening, I had been invited to a private party at the apartment of Mischa Groswald. It was a sumptuous meal where vodka and cognac flowed freely. My Canadian friends, Ross Mackay and Marie Sanderson, were present – a reassuring element – but also Vladimir Kotlyakov and Genady Golubev. Rumour in the West had it that Golubev was linked in some way to the KGB. In all, there were about twenty guests. Halfway through the meal, Grosswald (also a glaciologist) leaned across the table and said he hoped I was able to take good care of Mischa. I was astounded. Who else had heard? I answered

Yes, referring to Mischa (Mikhail) Vigdorchik, who had appeared on our doorstep in Boulder the previous year, in an effort to divert attention from the Plams. I explained that I had found him a temporary job and had been able to assist with the publication of his book *Arctic Pleistocene History and the Development of Submarine Permafrost*. In fact, I had sold him for one dollar our second car as we were leaving earlier that year for Switzerland and Moscow. To my amazement, Mischa Grosswald was not satisfied and enunciated more carefully, "No, I meant Mischa Plam. Didn't you meet him over the weekend?" I thought this was disaster. But there was no stir at the dinner table. Only later did Ross Mackay, always highly astute, ask for an explanation.

The next day was the penultimate day of the congress. Again nobody seemed to notice what was going on in my head. I was sure it must show on my face. However, Bruno was to depart for home the next day as he was very shortly to leave with his family for five month's sabbatical in Kenya. We agreed to have an early private dinner so we could discuss a wide variety of arrangements; for instance, I was to use his university office, and he gave me the keys to the family car and his comfortable chalet in the Canton Wallis.

We reserved a table for two in the very fine restaurant of Hotel Rossiya. We were placed close to the dance floor although as we sat down, few guests and no band had arrived. We were preoccupied with all our plans for the coming year and hardly noticed how quickly the restaurant filled. Strangely, the table next to us remained vacant. Eventually the small orchestra arrived and dancing began. We had long finished eating and were deep in conversation, but we couldn't help notice the arrival of two attractive women, finely gowned and groomed. They took the table next to us. It was impossible to avoid eye contact. Eventually we rose to leave. They in turn left their seats and came towards us as if they were expecting to dance with us. We gently fended them off as we left. KGB? We thought so, and it later became a family joke when Pauline explained to Beatrice Messerli that she would not let me go to the Soviet Union again unless Bruno was with me. Beatrice laughed-out-loud, as she had made a similar stipulation to Bruno, but in reverse.

Bruno left early the next morning and I took part in the closing assembly of the Congress. I had been re-elected as chair of the IGU mountain commission, with a strong level of East Bloc support. Again I was limousined to the airport, accompanied by the gracious Olga. I asked what literature she would like me to send as a thank-you present as she was still a university student and was studying English literature. She requested a pictorial edition of *Winnie-the-Pooh* (the first volume of A.A. Milne's famous children's tales of the 1920s) which, in due course, I had delivered to her.

At the airport, I was very surprised by the hasty arrival of Innokentiy himself who had come all the way out to say goodbye and hand me a present – a beautiful samovar (traditional Russian tea urn) and a quiet reminder of the invitation he had proffered in Boulder. The samovar is still proudly on display in my home. But this story is far from being played out.

Aftermath in the Swiss Alps

Throughout my sabbatical leave in Bern, there were repeated echoes of the previous summer's Soviet experience. In October, I attended the 30th anniversary of the British Glaciological Society in Cambridge. On entering the Scott Polar Research Institute immediately on arrival, I bumped into Terence Armstrong, who rushed forward to welcome me: "Jack, you are remarkably on time to join us for a glass of sherry. We are celebrating because we have just heard that Mischa and his family are safely in Vienna. So I joined with Terence, Hilda Richardson, Malcolm Mellor, and Gordon Robin, the institute director, to toast to Mischa's well-being and our success.[42]

The next two Soviet instances were less direct. First, a senior Swiss military officer was detected selling secrets to the KGB. This was a terrible shock for the Swiss as nothing like it had happened before. Another scandal followed close on the heels of the first, but of a very different nature. It transpired that a junior Swiss agent had been apprehended by the Viennese police for spying. To make matters worse, under interrogation he had pleaded that he was only practicing and that Vienna had been chosen because Austria was considered an "easy task" for an apprentice. He had been born in the Emmental (famous worldwide for its cheese). Europe was enjoying gales of laughter as the "story" was reported in many languages under the headline: "The Spy Who Came in from the Emmental" (LeCarré's book *The Spy Who Came In From The Cold* was still relatively recent at that time). The fourth incident was much more directly connected with mountain geography

and my 1976 visit to Moscow.

In March, shortly after the Messerlis returned from Kenya, Bruno received a telephone call from Genady Golubev. He and his family were in Vienna for a year (the "with family" was highly significant as only individuals with impeccable party credentials were allowed to leave the Soviet Union with their entire families) as he had a temporary appointment with the International Institute for Applied Systems Analysis (IIASA). Could Bruno invite him for a short visit to Bern and especially for an escorted tour of the tunnels that had been cut into the Jungfrau? The tunnels provided access to a Swiss military intelligence centre, although Genady explained that his interest was to inspect the glacier stratigraphy – the tunnels entered the mountain through the upper section of the Aletschgletscher and Genady was an internationally known glaciologist.

Bruno discussed Genady's request with me. It created a problem. Bruno did have official access to the Swiss military leadership, but the "Spy Who Came In From the Emmental" episode and the still burning top-level intelligence betrayal made for complications. Regardless, Bruno was able to obtain clearance and Genady arrived a few days later. The next day, accompanied by Hans Kienholz and my teenage son Tony, we went by car to Grindelwald and took the famous cog railway to Kleine Scheidegg and so onto the Jungfrau railway to the Joch. From there we first skied downslope, across OberMönchjoch and onto the Emigschneefeld where Hans Oeschger and a team of glaciologists were drilling to retrieve a long ice core for dating.

As we skied back to the Jungfraujoch,

Genady engaged me in casual conversation until I realized that his intent was to slow my progress and so separate me from Bruno, Hans, and Tony. When we were about 150 metres behind the other three, Genady stopped and indicated he wanted a strictly confidential conversation. This became for me one of those dream sequences one reads about in "real" spy novels. We were half-lost in light snow and patchy low cloud. The Mönch appeared and disappeared in the restricted atmosphere through which the sun sent occasional faint shafts. Bruno, Hans, and Tony were almost invisible, certainly beyond hearing. Was this real? Genady spoke, but even then in whispers: Greetings from Innokentiy and Rimma. My friends in Moscow were very pleased that Roger Barry, my close colleague who was serving as acting director for me during the sabbatical year, and I had taken care of Mischa in Boulder, Colorado. Now they could communicate with me in Russian, knowing that Mischa was a completely reliable translator. He could also ensure that my replies were in Russian. But it was imperative that Mischa's name was not included in any correspondence because everything would be inspected by the KGB. I thanked him, somewhat aghast and asked that he return my compliments and ardent thanks to Academician Gerasimov for last summer's hospitality.

We quickly caught up with our vanguard, skied beyond the Joch, and were allowed to enter the tunnels, the first sections of which had been cut through the Aletschgletscher. On the brink of a sharp turn, Genady called for a pause. He then gave us remarkable details of the stratigraphy that we were about to see round the blind corner. I sensed Bruno's

shock. What was Genady trying to demonstrate? He didn't explain, nor did we ask. Tony, newly turned 16, was flabbergasted: "Dad, whose side are you on?" I indicated my dismay – although a dark thought came into my mind: Was Mischa a KGB plant?

Thereafter, our family year in Switzerland continued and we were submerged in the tranquillity of the Emmental, which closely paralleled the rural atmosphere of the nineteenth century. My sabbatical research was divided between absorbing as much as possible from Swiss and Austrian experts in avalanche and related mountain hazards mapping and land-use planning implications and an attempt to determine the maximum thickness of Ice Age glaciation in the Bernese Oberland (Alps).

Mischa and his family were able to move from Vienna to Boulder, Colorado, and he took up residence at the mountain research station as its field director. His earlier experience in establishing the mountain station in the Caucasus proved highly relevant, and he played an important role in welcoming the surprise scientific delegation from China (Chapter 8). He also played a vital role in the 1980–81 establishment of the International Mountain Society and creation of our journal *Mountain Research and Development.*

The sabbatical year cemented the long-term relationship with Bruno. It also led us, unexpectedly, much farther into international mountain affairs following an invitation to lunch by Professor Walther Manshard (see Chapter 6). This was our first introduction to the existence of the United Nations University and proved crucial in leading us into the Himalaya, China, and eventually to the Rio Earth Summit in 1992.

Fig. 25.   Above the Grimselpass in the Swiss Alps with the Unteraargletscher in the centre. The Jungfrau, Mönch, and Eiger are barely visible on the horizon where the last episode of the previous summer's activities was played out (1977).

Fig. 26.   Dr Genardy Golubev (on right in green ski attire) visiting Professor Hans Oeschger's glacier deep drilling site (1977).

Fig. 27. The Jungfrau, queen of the Berner Oberland, in early morning light. The entrance to the tunnels (see page 66-67) are appropriately out of sight (1977).

Fig. 28. Appenberg in the Swiss Emmental, the 1990 birthplace of 'Mountain Agenda' and home of the Ives family during the 1976-1977 academic year.

Fig. 29.   The beauty and peacefulness of the interior of a Buddhist temple (1978).

CHAPTER 6

United Nations University
Agency with a Mountain Mission

I first learned of the existence of the United Nations University (UNU) on a spring morning during my sabbatical leave as guest professor at Bern University. Bruno came into my office early in May 1977 with the news that the two of us had been invited to lunch by Professor Walther Manshard at his home in Freiburg, southern Germany. I wondered aloud why the professor couldn't come to Bern. The number of days remaining for fieldwork in the Bernese Oberland was rapidly declining. Bruno protested and explained it would be courteous to accept and that Freiburg was only a pleasant morning's drive away. So I agreed, although somewhat reluctantly. I had not the slightest notion of the opportunity that was awaiting. It would have been a disaster to have refused.

After a convivial lunch, Professor Manshard told us that late the previous year he had been appointed as vice-rector of the recently created United Nations University headquartered in Tokyo. He and Frau Manshard would soon be heading for Japan, and that was why he wanted to be sure to discuss his plans with us before leaving home.

Then came a surprise. He showed us an organizational chart that he had recently put together. Under the university's new Programme on the Use and Management of Natural Resources in the Humid Tropics and Subtropics, a sub-programme was entitled "Ecological Basis for Rural Development in the Humid Tropics." This in turn had four areas of interest: (1) rural energy systems, (2) agro-forestry systems, (3) water–land interactive systems, and (4) highland–lowland interactive systems. The organizational chart gave details of the first three. The fourth stood out as a blank rectangle. We pondered the chart for a few minutes before the big question: Would either of us be prepared to take on the role of 'coordinator' for Project 4?

UNU's Highland–Lowland Interactive Systems Project

Bruno explained that he was a member of a small university institute and he felt there was no way that he could take on any such responsibility. I was thinking very hard: what would I be letting myself in for?

J. D. Ives, *Sustainable Mountain Development*, https://doi.org/10.1007/978-3-030-96029-2_6

Manshard waited. I asked him to explain what was meant by "highland–lowland interactive systems" and how the blank space was to be filled. He replied that would be entirely up to me. My interest was sparked, so I put the next leading question, "How much was committed for the budget?" He replied that he could guarantee a minimum of US$100,000 per annum for three years, perhaps more, although there was no provision for salary, it being considered an entirely *pro bono* appointment. In those days $100,000 per year was a highly significant level of funding, especially considering that it didn't entail provision for any salary or consulting fees. I thought further and answered that I would take it on, provided we could start in the Himalaya to tackle the problem of the causes and effects of deforestation and related mountain hazards. He agreed to that, but only if I would first undertake a mission to Chiang Mai, Thailand, and help determine whether or not UNU could make any impact on the problems that were being created by opium production amongst the "hill tribes" and its attendant environmental and social damage. The project would involve collaboration with Chiang Mai University, so a visit would first be required to assess the faculty's competence and commitment to the outline plan. I took a deep breath and asked when we could begin. He proposed that I come to Tokyo the following April (1978) and go to Thailand with Professor Gerardo Budowski in late April/early May.[43] At that moment, a fundamental decision had been made.

A toast to future collaboration was offered in white Rhine wine. Then I asked how did Professor Manshard come to invite me to lunch. Walther, as we soon addressed him, despite German academic propriety, replied that it was partly due to an anonymous source whom he had met the previous week at a UNESCO conference in Reykjavik. I told him that was not sufficient, so he opened up: "Your countryman, Jim Harrison." I took this as a real compliment from the former director of the Geological Survey of Canada with whom I had fought tenaciously through the 1960s about the status of geography in the Canadian federal government – although he had already written a key reference for my Guggenheim award. As he was chair of the newly created committee that would oversee Walther's entire UNU programme, I would be meeting him in Tokyo the following spring and working under his general supervision. But what did the name of the new project – "Highland–Lowland Interactive Systems" – imply? I thought it was sufficiently akin to "mountain geoecology"; Walther smiled agreement, and that was how it began. But Thailand! I was uncomfortable over the thought of a hot humid climate, my preference being north of the Arctic treeline or above the alpine treeline. A joking admonishment for my geographical inadequacy – surely I was aware that there were mountains in northern Thailand?

Bruno and I drove back to Bern through the late afternoon and evening. Over dinner in the Jura, we talked extensively about what Walther would lead me into. As things turned out, within eighteen months, Bruno had joined me as "co-coordinator" of Project 4.

A New Type of UN Institution

The United Nations University was established in 1973 as the brainchild of former UN Secretary-General U Thant and became active in 1975. The programme on the "Use and Management of Natural Resources in the Humid Tropics and Subtropics" was set up in 1976 with Walther Manshard's appointment as the third of three UNU vice-rectors, hence the hastily arranged meeting in Freiburg in May 1977. The University's raison d'être, in short form, is stated this way:

> *The United Nations University is an academic arm of the UN system implementing research and education programmes in the area of sustainable development with the aim of assisting developing countries.*

It was not to be a conventional university; rather a form of think tank for the UN system. Its chief administrator, the rector, also carried the rank of under-secretary-general of the UN, reporting both to the secretary-general and to the director-general of UNESCO. The university had an advisory council of 24, also appointed by the secretary-general and the UNESCO director-general.

Financially, it depended entirely on direct annual contributions from those UN member countries who could be persuaded to assist; thus, it had a relatively modest operating budget with the danger of year-to-year fluctuations. It also had a remarkable degree of independence as a basic research and educational institution and its staff and academic associates were free to analyse development projects relevant to their research, regardless of the sponsor, as long as any critical assessment was based on solid scholarly investigation, as opposed to political opinion. This factor was especially important to me and the highland–lowland interactive systems project as we got underway with our research. For the most part, we were largely free of the need for "political correctness." Academic independence is critical for any research undertaking that may produce controversial results, and in the 1970s and 1980s our results were certainly controversial. As we increasingly questioned the assumption of Himalayan deforestation catastrophe our results were more and more in conflict with predictions publicized by the World Bank, the Asian Development Bank, various aid agencies, and government officials.

From the beginning, the major financial contributor to UNU was Japan, and for this reason, the university's headquarters were in Tokyo. In the early years, the headquarters were located on the 15th and 16th floors of an impressive building in Tokyo's Shibuya district, affording a distant view of Mount Fujiyama. The high rental costs were covered by the Japanese government until subsequently a superb new building was erected a short distance away – no doubt some of the most expensive real estate in the world.

My involvement as a project co-ordinator began in the very early years of the university's existence. Certainly, between 1977 and 1992, major decisions, including initiation of new projects, were made by senior administrators, especially the vice-rectors, subject only to general overview by the rector and governing council. Thus it was that

Walther Manshard was able to recruit me on his own initiative, subject to formal confirmation. He was also able to provide me with great latitude to frame the project, subsequently in close conjunction with Bruno. I do not recall a single instance when requesting budget increases, appointment of additional post-doctoral fellows, special workshops, even long-term financial support for our new quarterly journal *Mountain Research and Development*, that did not receive his immediate approval, sometimes on the basis of only a long-distance telephone call.

Perhaps UNU's major difference from a conventional university, at least in its first two decades, was that it had no permanent academic staff, nor students in formal courses. The "academic staff" as such were enlisted as unsalaried associates on the understanding that they were effectively on loan from their regular institutions. Consequently the home institutions of some of the more central of the associates were designated as UNU affiliate institutions with formal contracts signed by the senior administrators of the respective units. In the early years, the universities of Bern, Colorado, Chiang Mai, and the Australian National University were among the earliest affiliates under Walther Manshard's general purview. Harvard was the only other U.S. university affiliate in addition to Colorado at the time.

The foregoing comments, of course, refer to the early years of UNU's operation, a period when maximum flexibility could be expected, although I was aware from that lunchtime discussion in May 1977 that Walther was a very dedicated person with whom it was a joy to work. Much of our rapid progress depended heavily on Walther's creative approach to management and administration.

Another significant benefit of the early years was that the project co-ordinators were invited to headquarters twice yearly to meet with Walther, with the members of the scientific advisory committee, and with each other, to discuss progress and debate future plans. A great sense of camaraderie and of loyalty to UNU was created by these meetings.

During the last 20 years, when my direct involvement has become slight, UNU has expanded significantly. Major UNU specialized research centres have been established in several countries, including Finland, Canada, Germany, West Africa, and Costa Rica. In recent years, the university has created a master's degree programme and a wide range of projects that explore and apply the most recent advances in research and educational technology.

As a quick follow-up to the lunch meeting in Freiburg of May 1977, Walther Manshard visited me the following October in Boulder, Colorado, en route from his home in Germany to Tokyo. I had already received a confirmation of my UNU appointment as co-ordinator, Highland-Lowland Interactive Systems Project. After an afternoon walk-around and introduction to the graduate dean on the main campus, I arranged a dinner, to include all INSTAAR faculty and staff who were available. Walther gave an enthusiastic after-dinner talk outlining what he thought would be "an appropriate expansion of INSTAAR's already impressive activities." He also announced that I had been invited to Tokyo the

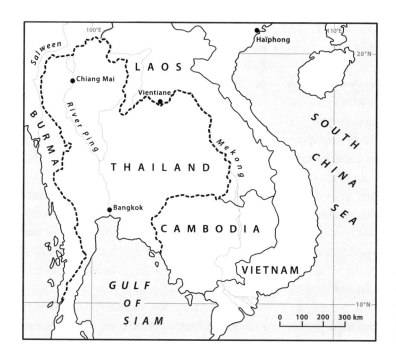

Map 2. Outline map of Southeast Asia emphasizing Thailand with surrounding countries and showing the locations of Bangkok and Chiang Mai.

following April, and from there to northern Thailand and the Himalaya.

I arrived in Tokyo in mid-April 1978. It was cherry blossom time, and the city parks and the Imperial Palace grounds were gorgeous. I also recall that UNU headquarters, high above the Shibuya district, had on display numerous and appropriate notices informing staff and visitors how to behave in the event of an earthquake.

Under the chairmanship of Jim Harrison, my old Ottawa adversary, who now served as assistant director-general of UNESCO, I met my other project co-ordinator colleagues. Geographers were dominant. Jack Mabbutt (Cambridge/Australia and co-ordinator for an arid lands project) was particularly effective in the general discussions that followed. Another valuable participant was Ingvar Friðleifsson,

who was coordinator of an Icelandic project for introducing Third World scientists and students to advanced geothermal power development. One of my most important new colleagues, however, was Gerardo Budowski[43], coordinator of the agro-forestry project, with whom I was destined to work in Thailand.

The main purpose of bringing all the project co-ordinators together under the UNU umbrella in Tokyo was for us to get to know each other. We were also able to discuss with Walther and Jim Harrison the approaches to office management and field reporting – a valuable exercise in creating dedicated team rapport. After a lively discussion and lunch in the Japanese restaurant, I met with Jim, Walther, and Gerardo, as Gerardo and I were to depart for Bangkok and Chiang Mai, Thailand's old capital city, the following morning.

My First UNU Mission: Chiang Mai and Opium Poppy Cultivation, 1978

Chiang Mai University had approached UNU just over a year previously with a proposal for some form of affiliation and cooperative research. In these early days of its evolution, it was important for UNU to form liaisons with Third World universities whenever there was reasonable confidence that they would be relevant to the overall UNU mission and would include a significant element of applied research contributing to policy development. Hence, my preliminary visit to Chiang Mai University with Gerardo carried a heavy responsibility, particularly as I was transparently a novice in these undertakings. In this I depended on the much more experienced Gerardo.

Dr Pisit Voraurai, agronomist and vice-rector of Chiang Mai University, had spent several years working with the hill tribes in northwestern Thailand (principally the Karen, Lisu, and Hmong) in an effort to introduce alternative farming practices to opium poppy cultivation. He had developed a significant research station in the mountains about 60 km north of the town at Huai Thung Choa where agricultural and socio-economic research was being carried out. This would be the most important facility for Gerardo and me to investigate. In addition to Pisit's research team, German, Australian, and USA/Hawaiian groups were engaged in similar work.

UNU was particularly interested in a proposal that Chiang Mai University could cooperate in applied research that might lead to solution of environmental, socio-economic,

and political problems. These had become increasingly pressing throughout Southeast Asia since the end of World War II, and especially following the Vietnam War. In this respect, Pisit's work at Huai Thung Choa was central to our mission.

Chiang Mai lies in the valley of the Ping River about 600 km north of Bangkok and is surrounded by forested mountainous terrain that rises to between 1500 and 2000 metres above sea level. In 1978, the region was still awaiting discovery by mass tourism. This was fortunate for us as the old town remained very attractive and little distorted by modern development.

The Chiang Mai region can be divided into two distinct landscapes: the extensive north-to-south lowland that constitutes the broad valley of the River Ping, one of Thailand's major rivers, and the rugged mountains (often referred to as hills) that are heavily forested. This physical division is mirrored by a pronounced ethnic, socio-economic divide. The rich agricultural lowland is densely populated by ethnic Northern Thai (Khon Muang), who are intensive rice cultivators. The forested mountains are the home of several distinct hill tribes. The most numerous in the Chiang Mai-Huai Thung Choa region are the Karen (who were partially assimilated into the lowland Thai system), Lisu, Lahu, Akha, and Hmong – in general occupying progressively higher land. Although the Karen had almost entirely abandoned opium poppy cultivation, the other groups depended heavily on it, and the entire region was part of the so-called golden triangle of the time that embraced a large section of the highlands from Vietnam in the east to Burma and northeast India in the west, and northwards into Yunnan, China. The patterns of agricultural production have been referred to traditionally in Western text books as "slash-and-burn agriculture." This is a pejorative term, and it is recommended that it be replaced with the more technical term "swidden agriculture."

In simple terms, swidden entails the controlled burning of small patches of forest to enable the cultivation of poppies during the winter dry season, alternating with maize and hill rice, depending upon altitude, during the summer monsoon. The poppy cultivation quickly exhausts the soil and is usually practiced for only a few seasons, after which the plots are abandoned to fallow so that the forest cover re-establishes itself. Numerous other food crops are grown, and each village has several plots in various stages of development. The key to sustainability of this system (or systems) is the length of forest fallow because the ash from the controlled burns provides the main source of fertilizer. This in turn depends upon the size of the population per unit area that needs to be fed. A rapidly growing population following World War II had

Fig. 30 [top].   Chiang Mai University students of traditional Thai dance made a special performance for the camera. Note the hand of the instructor who is making a small correction of position (1979).

Fig. 31 [bottom].   On arrival in Chiang Mai, northern Thailand, we learned that the white poppy was promoted as the best for producing high quality opium, 1978.

Fig. 32. Late afternoon light looking over the hilltops above Huai Thung Choa towards the frontier with Burma (Myanmar) (1978).

Fig. 33. The Huai Thung Choa experimental gardens. Sufficient progress had been made so that cut flowers, seeds, and some vegetables were already being sold in the markets of Bangkok, Singapore, and Hong Kong (1978).

Fig. 34. Visit to a traditional Lisu village within walking distance of Huai Thung Choa. It is May, the hottest month of the year, and roast termites are being prepared for lunch in the cool half-light of the headman's house (1978).

Fig. 35. The Lisu swidden fields, while small in size, are planted with dozens of different food crops, although in this particular case the dominant crop is hill rice (1979).

forced reductions in the length of the period of fallow. In response, forest was being replaced by Imperata grassland, regarded as a much less productive cover type, if not an invasive weed, extremely difficult to eradicate. Steep slopes, increases in population, reduced periods of fallow, spread of *Imperata* grassland, and accentuated soil erosion were all contributing to perceived serious socio-economic and environmental problems. International efforts to eradicate opium production pervaded the entire scene.

The Thai forest service, aided by the military, were reforesting the invasive grasslands with plantations of native pine, both to reduce the effects of soil erosion and as part of its overall strategy to control the opium trade. However, pine forests were regarded by the hill tribes as a serious threat to agricultural productivity. Consequently, they frequently resorted to surreptitious forest fires to eliminate the hated pine plantations. A final complication was a significant movement of hill tribal people into the northern Thai hills, a traditional occurrence across the unmarked frontiers of the region that had been accentuated by the Vietnam War.

It was into this morass that Pisit and the several outside agencies and research organizations had entangled themselves. Replacement of opium poppy cultivation with politically more acceptable cash and subsistence crops on terraced slopes (terraces were not part of the traditional swidden agriculture) would seem an obvious solution. One very serious difficulty, however, was that, legally, the hill tribal people held no status as Thai citizens. There was strong Thai-hill tribe prejudice, and the tribal people had little legal defence if their traditional land was appropriated.

On our first day in the Chiang Mai region, we were given an extensive tour of the campus and talked with many members of the faculty and graduate students, as well as some of the leaders of the Australian, U.S., and German projects. Early the next morning, we departed for Pisit's research station at Huai Thung Choa. It was very hot – May is Thailand's hottest month of the year – and I was pleased when we left the valley of the Ping and headed up the heavily forested hills to higher ground in the general direction of the Burmese frontier. The four-wheel-drive vehicle was a good choice for conveyance along the bulldozed mountain road that reaches an altitude of close to 1800 m although Huai Thung Choa itself is situated at about 1300 m.

We spent only three nights at the research station, but Gerardo and I were impressed with everything we saw. Pisit had succeeded in creating a school staffed by Northern Thai volunteers who had the challenging task of teaching children from all the local ethnic groups, several of whom had a history of inter-racial violence and competitive involvement in the opium trade. His principal effort was to introduce the growing of vegetables, cut flowers, and garden flower seed production for sale in Chiang Mai, Bangkok, Hong Kong, even Tokyo, as alternatives to poppy cultivation.

The flower and vegetable gardens were impressive but, as Pisit admitted, success depended on a number of critical issues – transport to market, even refrigerated transport;

sustainability of cultivation practices, especially related to soil erosion; and ability to induce local minorities to accept the arduous task of terracing steep hillsides and to adopt other soil conservation measures. In all of this, the effectiveness of poppy cultivation was a serious challenge. Poppies provided a light, high value, easily transported product that surreptitious traders collected directly from the villagers. Taking the raw product, they disappeared into the jungles along the Thai–Burma border and, after processing, took their "value-added" opium to waiting boats on the Andaman Sea. The poppy provided the cash crop for the ethnic highland villagers while other agricultural activities ensured virtual self-sufficiency in food production.

A highlight of our visit to Huai Thung Choa was a long morning walk off the road to one of the more isolated Lisu villages where Pisit had persuaded the people to attempt an alternate cash-cropping system. We looked down through the forest onto a cluster of about 30 thatch and bamboo houses, then walked down to be received by the village headman and invited to lunch.

This was my first "real" experience of being entertained in an isolated indigenous village. We took off our boots, entered the headman's hut, and were seated on thatch mats. Then in the half-light, there being no windows, we realized how cool it was in contrast to the heat outdoors. Small dishes were gracefully set before us and as Gerardo began to eat without hesitation, I followed suit. A strange dry crunchiness prompted me to whisper to Pisit, "What are we eating?" That was a mistake. The answer – roast termites! Even in the

half-light, Pisit sensed that I was somewhat taken aback. So he spoke quietly to our host. My plate was removed and replaced by another – quite different. Lunch proceeded and it was most appetizing. But when I asked Pisit what I had eaten, he replied that he would explain after we had left. It turned out that the termites had been replaced with a whole range of different roast insects. Since that time, I have eaten, and mostly enjoyed, everything set before me in any part of the world, except for Tibetan salt tea.

After saying our thanks and goodbyes, we wended our way back to Chiang Mai. It was Thai New Year's Eve and the rector had invited us all, together with an assortment of deans and senior faculty, to a feast. We assembled in an impressive traditional Thai building set in an iron-gated courtyard. There were many courses, accompanied by several alcoholic beverages, the obvious favourite being what I could best describe as bamboo schnapps. The rector was especially partial to it. Gerardo and I were sitting on either side of him. As the dinner was coming to a close, he turned to me and asked if I would like to complete the evening in traditional Thai New Year style. I concurred with thanks, too late to see Gerardo's serious frown. So we walked down to the courtyard, the rector already unsteady on his feet. Awaiting us was the rector's brand new white Mercedes-Benz, delivered only the week before. Pisit offered to drive it for him and received a sharp rebuff. Gerardo managed to warn me that we must be very careful as Pisit could even lose his job if events took a turn for the worst and the rector felt insulted. As he spun around his new white car to drive through the iron gates, he casually scraped the passenger side from end to end.

We were driven into the centre of Chiang Mai as our tension mounted. The streets were teeming with people wearing costumes of all kinds. It appeared that all the hill tribes had descended into town, and there was a melee of laden bullock carts and small children everywhere. As we approached the town's central square, the press of the crowd became impossible. Eventually we were forced to stop. I was amazed to see that the rector proposed to abandon the car and walk to a gaily lit and festooned building a short distance away. In the ensuing confusion, Gerardo managed to speak to me privately: "Jack, a very serious question. Do you want to go through with this?" Gerardo's explanation made me realize that I was extremely tired, so we expressed our regrets and found our way back to our hotel.

We reported to Walther Manshard by telephone from Bangkok airport that we thought Chiang Mai University would be highly suitable for UNU affiliate status and that the Huai Thung Choa experiment was worthy of substantial expansion. However, we suggested that, for good measure, a full workshop should be scheduled for Chiang Mai as soon as possible and that other researchers, experienced in the northern hills, should be invited. Walther immediately agreed and asked for a detailed proposal and timetable.

I was able to communicate with Gerardo and several other vital players, and UNU scheduled a formal workshop for the following November in Chiang Mai. A research plan was then submitted, drawing heavily on the experience of the workshop.

Return to Chiang Mai: Next Steps towards a UNU Project

The first UNU mountain workshop in Thailand was held 13–17 November 1978 in Chiang Mai. Walther Manshard himself chaired the meeting and our host was Professor Pradit Wichaiyadit, rector of Chiang Mai University. HRH Prince Bhisatej Rajani, director of the Royal Hill Tribes Development Project, delivered a gracious welcome.

We had managed to attract many of the leading figures in hill tribe research, both Thai and foreign. Well known among them was Professor Sanga Sabhasri, secretary-general of the Thai National Science Foundation.[44] I also arranged for Kamal Shrestha, our Kathmandu manager for the Nepal project, to be invited. The participants included virtually all the key players providing important links with several of the major institutions, such as the East-West Center of the University of Hawaii, and the Australian National University. Fifteen research papers were presented and discussed and written summaries of six panel discussions were submitted.[45]

Pisit's work at Huai Thung Choa, together with the highly relevant proceedings of the workshop, laid an excellent foundation for our proposed research plan, and his rapport with many of the local people and their leaders was a vital asset. The entire group endorsed an approach that would try to determine if and how we could modify the way of life of villages representative of three different ethnic groups.

The basic aim of transforming people and landscape from a swidden, opium- producing microcosm of the hills of northern

Thailand to one of sedentary, terraced agriculture geared to produce legal cash crops was an enormous challenge, and several of us worried more than a little about the moral justification of our efforts. What we were attempting, if successful, would be infinitely more humane than an earlier CIA plan of handing out cash to the swiddening people and expecting them to cut back on poppy cultivation. In those early days, twinges of conscience were rather easily subdued; the extent of our anthropological understanding was conspicuously inadequate.

We agreed to focus on a select number of Karen, Lisu, and Hmong villages located along an approximate altitudinal transect above and below Huai Thung Choa. It would entail field testing a number of proposals derived from the workshop: selection of a range of suitable crops as alternatives to opium poppy; determination of amounts of soil loss among these crops in comparison with losses under maize or opium production on steep slopes; attempts to stabilize production and maximize yields by construction of terraces; ethnological investigation of the different tribal groups and their inter-relations; and

Fig. 37.   A young Lisu woman working at her spinning, Huai Thung Choa area (1979).

Fig. 36.   A special friend. A Lisu lady who, on my several visits, always handed me something new to eat from the nearby swidden. She is chewing betel (1979).

analysis of market opportunities and the necessary transportation links.

Clearly, we were preparing a complex multidisciplinary undertaking. A parallel study would be the environmental and socio-economic impacts of ethnic Northern Thai (Khon Muang) lowland sedentary farmers, dependent on intensive irrigated rice, who were moving upslope into hill tribe territory. Not only were they practicing a form of careless and intermittent swiddening, but illegal logging was also a complicating factor. Their Majesties, the King and Queen were respected as the Father and Mother of the hill tribes, so it appeared at the outset that royal involvement should somehow be sought. In effect this happened immediately; obviously our senior Thai colleagues were in communication with the palace.

The Geographical Institute in Bern, under the field leadership of Dr Hans Hurni, had made great progress with applied agricultural and soil erosion research in Ethiopia. It had involved setting up large soil erosion study plots that seemed highly relevant to the UNU project in northern Thailand. With Bruno's encouragement, Hans expressed enthusiasm to apply his extensive experience to the Huai Thung Choa area. Under Hans's direction, large soil erosion study plots were set up and instrumented for testing for relative soil losses amongst various proposed cash crops in comparison with the traditional maize and poppy dominant. In effect, we became producers of opium, albeit on a very small scale. The study also involved determination of how any proposed changes in village cropping patterns, together with the associated and necessary adaptations to a different form of livelihood, could be attained and made sustainable. Hans began field operations during the late summer of 1979 with a group of young Thai research assistants.

One of the tangential results of publishing the proceedings of the 1978 Chiang Mai workshop arose from Walther Manshard's impatience over the time needed by the newly established UNU Press in Tokyo. This prompted an entirely entrepreneurial response from me that led to a major breakthrough for the mountain cause. I suggested that it should be possible to complete the publication process in half the time and at half the cost. The foundation of the International Mountain Society (IMS) as the device used to provide a framework for publication of the quarterly journal *Mountain Research and Development* was the result. This was possible because Walther arranged for UNU to provide an annual grant of US$18,000. Together with normal individual and institutional subscriptions and guarantee of one-time funding from UNESCO and UNEP for the publication of specific issues, the creation of the journal was assured. It was a joint IMS-UNU publication and has become the prime international mountain quarterly. My wife Pauline and I served as editors and managed to produce four issues a year for nearly 20 years. Pauline carried the lion's share of the workload. The IMS, collectively with the UNU, the IGU mountain commission, and our more informal group of mountain advocates thereby obtained a powerful vehicle to support the mountain cause that greatly assisted the eventual achievement at the Rio

Earth Summit in 1992. I was finally able to take Klaus Lampe's 100 DM from my wallet and put it to its proper use.[46]

A second, much more extensive workshop was held in Chiang Mai in November 1979. The proceedings, which already included some of the first definite field results of Hans Hurni's work in the Huai Thung Choa area, were published in the new journal in May 1982.

A Surprise Royal Invitation to Dinner

In early January 1979, I received an urgent telephone call from Walther Manshard in Tokyo: their Majesties King Bhumibol and Queen Sirikit had requested my presence at a small, informal dinner party on 17 February at the Winter Palace above Chiang Mai. Was I free to attend? I did not pause before assuring Walther that I would be there. He explained that this would be a first-class occasion for me to attempt a discrete opening to raise the question of a financial contribution to UNU. Regardless, it augured very well for the UNU/Chiang Mai collaboration. Walther assured me that UNU would cover all my expenses, so I immediately accepted the invitation to the most expensive dinner of my life (that is, expensive in terms of travel costs).

Determined to use the occasion to enhance planning for our Huai Thung Choa project, I arrived in Chiang Mai a week early. Much of 12 February was spent with Pisit and University Rector Tawan in research planning and meetings with other university colleagues. Early the following day, Pisit drove me in the newly acquired UNU Land Rover to Huai Thung Choa to await arrival of King Bhumibol and a large entourage to inspect the field station. I was impressed that the King drove himself and, despite being quite close to the Burmese frontier, had no significant military escort. In fact, as his jeep drew up and he jumped out of the driver's seat, I all but made one of the giant gaffes of my life by mistaking him for an employed driver.

The royal visit appeared highly successful. Pisit was more than satisfied. Following the royal inspection of the field station, I walked extensively with a small group of Thai colleagues and visited several Lisu and Hmong villages. Back in Chiang Mai on the Saturday, I interviewed three Thai graduate students whom Pisit had encouraged to apply for UNU fellowships, then spent most of the afternoon being primed by Pisit and Rector Tawan for the evening's important dinner engagement.

The Winter Palace is situated in the hills high above Chiang Mai. With Pisit and the rector, I was driven up a well-paved scenic road in the gathering dusk. We were met by gracious palace attendants and ushered into a beautifully decorated anteroom to await arrival of Their Majesties. Here I learned that I was to be accompanied to dinner by three other guests, one of them a Brit of the "old school" tradition who was a senior FAO official.

As we awaited the appearance of Their Majesties, the attendants subjected us to some mild teasing. For instance, a statement that His Majesty took very seriously his grandfather's protocol (as in the movie *Anna and the King of Siam*) – so for our own safety, each of

Fig. 38.   A royal visit to Huai Thung Choa. His Majesty King Bhumibol is viewing the gardens from a specially prepared shelter. Besides him, in military dress, Dr Pisit Voraurai demonstrates the success of his venture (1979).

us should make sure that our head was below the level of His Majesty's or else we may lose it; his sword was very sharp! I remarked that I thought the movie was banned in Thailand and was told to my amusement that, yes, in public, it was banned, but it was nevertheless one of the King's favourites.

The King entered down a short flight of stairs to greet us. He was not wearing his grandfather's royal robes and, after the formal Thai greeting of hands pressed together and bowed heads, shook hands informally and welcomed us. He did not even sport a dress sword! Then he turned to hand Queen Sirikit down the stairs and we were presented. She was splendidly gowned and bejewelled. I was thrilled to be introduced to the most beautiful woman I have ever met. She was radiant.

After brief conversations we were ushered into a resplendent dining room leaving Pisit and the rector to be entertained by the court officials. We were seated by colourful uniformed attendants, two-by-two facing each other. Queen Sirikit sat at one end of the table; King Bhumibol took the opposite end on my immediate right. Aperitifs were poured and the King turned immediately to me and began a long serious interrogation about Huai Thung Choa and the application of remote sensing. I had been warned that he was highly educated, sharply astute, and very serious. He explained that he was well aware that Dr Voraurai was hoping for his support to obtain low-level vertical colour photography for the field area and surroundings. However, only the week before, he had received a distinguished group of NASA officials who explained that the new LANDSAT imagery was capable of obtaining minute ground detail, so why employ low-flying aircraft? I summoned up the courage to demur based on my own experience in the Colorado Rockies with LANDSAT imagery, noticing that my FAO Brit companion was becoming somewhat aghast as I differed with the King.

Eventually Queen Sirikit interjected on my behalf. She used a form of parable – "My dear," she addressed the King, "you remember how you always insist that we must visit hill tribe villages using our helicopter to make personal contact in their villages, and how I have to go, despite being fully gowned. Surely Professor Ives is arguing the need for the same approach – close contact." I was amazed that she had interceded on my behalf. Next, she turned to me and said, "Is it true what my spies tell me, that the other day you walked more than twenty kilometres through the jungle?" She followed up by asking my age. Very much taken aback, I confirmed the rumour and stated my age as 48. I was further taken aback when she expressed her admiration, saying that she was the same age. At this point and in a state of confusion, I insisted that I must be at least ten years older than her. She didn't miss a second and replied, "Well, on your next visit I will ask you to undertake that feat again and I will follow you in the royal helicopter to check out the distance." Now, without being able to think clearly I replied, "Your Majesty, if you will do that, I will make it thirty kilometres." At this point my FOA colleague appeared on the brink of apoplexy – it was bad enough for me to argue with the King, but to risk flirting with the Queen was unpardonable. The King gave a hearty laugh and diverted the conversation back to remote sensing while a second round of aperitifs were being poured.

The Queen intervened once again: "Gentlemen, if you are to avoid going hungry, you must not wait for my husband to begin as he can be much too serious at times." And so began a royal dinner with vigorous informal conversation and much camaraderie. It lasted until almost midnight, an experience never to be forgotten.

King Bhumibol demonstrated an enquiring mind, highly perceptive and intense. He let me see how anxious he was for the welfare of the mountain minorities and strongly welcomed the growing collaboration between UNU and Chiang Mai University; Queen Sirikit was closely in step.

The invitation to the winter palace above

Fig. 39. Dr Hans Hurni's, large soil erosion/alternate cash crop study plots in the Huai Thung Choa area. One of the plots was used to grow an opium poppy-maize rotation so that the amount of soil loss between the different crops, including the traditional form, could be compared (1980).

Chiang Mai had proved a fascinating experience for me, and I think UNU was well repaid for covering my travel expenses. Although I claim no direct connection, somewhat later UNU received a five million dollar grant from Thailand. The report I subsequently submitted to Walther Manshard became part of the basis for serious development of the Huai Thung Choa project. Three Thai graduate students spent eight weeks at INSTAAR's mountain station the following summer, together with four students from Kathmandu, all on UNU fellowships.

Evolving Issues and Critical Problems

By the end of 1980, the collaboration with Pisit Voraurai and his Chiang Mai colleagues was flourishing. Early results from Hans Hurni's field investigations, enhanced familiarity with the local people, and the accumulation and publication of the results of the two Chiang Mai workshops had created a significant knowledge base and a broader understanding. Most important was awareness that the early assumptions we carried with us on our first arrival in April 1978 needed profound adjustment. In less than two years, a number of challenges to conventional wisdom were brought to the surface, initially cloaked as hypotheses:

- Swidden agriculture was not the demon of environmental destruction that had been widely assumed.
- Spreading *Imperata* grassland as replacement for natural forest was not entirely bad and was not nearly as difficult to eradicate (or tolerate) as previously thought.
- Replacement of *Imperata* grassland with forest plantations using *Pinus keysia* as the principal seed source (an approach preferred by the Royal

Thai Department of Forestry) was counter-productive.

- The ethnic Northern Thai (Khon Muang), together with commercial logging (legal or illegal), accounted for more environmental degradation than could be achieved by all the hill tribes combined.
- A new approach to "development" among the hill tribes was needed.

These five points made us recognize that we had disturbed a hornet's nest. As mentioned earlier, the pejorative term "slash-and-burn" seemed largely a Western construct that was inherently damaging and was one of many ethnic prejudices that quickly surfaced. Given adequate length of forest fallow, the many swidden systems deployed by the various hill tribes were essentially sustainable (see Kunstadter et al. 1978; Forsyth 1996, 1998; and Rerkasem 1996 for a much more extensive analysis, based in part on our earlier findings).

Hans Hurni's field investigations led to the conclusion that the practical, or purely technical aspects of intensification of agriculture, management of soil erosion, and application of cash crops as alternative to the poppy could be resolved. The underlying problem was a social one. To develop terraced agriculture, even if the resilient swidden patterns could be modified or restricted, required a larger labour force than was locally available. Already, the labour requirement of the experimental efforts was attracting inmigration by ethnic Northern Thai from the overpopulated Ping Valley lowland. From this it appeared that the land productivity versus population ratio would likely prove an extremely difficult issue to resolve.

Nevertheless, this first UNU highland-lowland interactive systems mission produced a large body of relevant research results and a much fuller understanding of the problems facing this small section of the Golden Triangle. The overall socio-economic situation began to change rapidly, however, as the tourist juggernaut got underway in the late 1970s and 1980s. The villages of the hill tribes became international tourist attractions, providing new sources of income but also the negative aspects of outside disruption. The City of Chiang Mai became a major tourist destination. However, it would be too facile to imply that international tourism would "solve" the "problems.'" The apparent reluctance of the Thai government to grant the hill tribes full rights of citizenship remained a serious obstacle to balanced development and environmental management (Rerkasem 1996).

Fig. 40. The Kakani ridge, a short distance from Kathmandu, was selected as the Middle Mountain intensive study area. The forest to the right of centre is a religious site and hence protected from tree-cutting. The main crest of the High Himal shows white on the skyline. The Trisuli Road twists and turns through the centre.

CHAPTER 7

Himalayan Reconnaissance 1978
Deforestation, Landslides, and Farmers

I expressed my deep gratitude and bade goodbye to Gerardo Budowski on the Bangkok airport following our visit to Chiang Mai and northern Thailand and departed for Darjeeling, Kathmandu, and Dehra Dun (central Indian Himalaya about 400 km almost due north of New Delhi). This was the second stage of my highland-lowland commission: to select a possible field area for investigation of what Erik Eckholm's effective journalism had turned into one of the great environmental truisms of the day – that the Himalaya were on the brink of losing all their forests and topsoil and, as an outcome, to spread death, destruction, and catastrophic flooding in Gangetic India and Bangladesh. My general objective was to locate suitable field research areas where we could evaluate this staggering claim. Ideally, the area selected should be represented by an altitudinal transect from the floodplain of the Ganges, through the Middle Himalayan Mountains, to the High Himal. There should be comparative ease of access to the full range of Himalayan altitudinal vegetation belts, to peasant villages embracing several ethnic groups and, of critical importance, no political restrictions.

On arrival at Bagdogra airport, to my delight, Yuden (see Chapter 4), escorted by her much older brother, Norbu, was waiting for me. Yuden, at the time, must have been nine. She was dressed in her traditional costume, and it was obvious that the heat of Bagdogra-Siliguri at low altitude was oppressive. Norbu explained that she had never been so close to sea level before. Understandably, we were glad to reach higher altitude quickly in the taxi that I hired to take us up the famous mountain road, criss-crossed by its equally famous narrow-gauge railway line that I had first taken in 1968 with Fritz Müller and Barry Bishop. Norbu and Yuden left me at the entrance to the Everest View Hotel with a promise that they would see me the next day.

The hotel was a conspicuous and impressive relic of the British Raj, and to be consistently addressed as "Yes, sahib; no sahib; this way please, sahib" was a little disconcerting. However, "Would sahib like to be awakened early to see the sunrise on the Himalaya?" met with a firm "Yes, please."

I had dinner the first evening with Dr Dilip Dey, whom I had met in Kathmandu during the 1975 MAB-6 workshop, and several of

J. D. Ives, *Sustainable Mountain Development*, https://doi.org/10.1007/978-3-030-96029-2_7

his colleagues. All were very supportive when I outlined the drift of my UNU commission that aimed to examine the causes and timing of deforestation, with mountain hazards mapping as a primary component. However, they warned me that, while physical access to a full mountain transect should present no serious difficulty, politically I would face implacable opposition from the Indian authorities, and especially the military. Any field area that would meet the requirements of the proposed research would need access to the restricted military zone that included most of Sikkim. In other words, a research team would never get near any of the glaciers, let alone the frontier with China (Tibetan Autonomous Region). They advised that Nepal would be much more amenable.

Early the next morning, I was gently awakened in the dark. On arising I was robed and guided to the private balcony where a small table was set for morning tea with Arrowroot biscuits in the approaching dawn. There I sat, like some imperial dignitary, sipping the mandatory Darjeeling tea and watching in wonder as the eastern sky brightened and the first rays fell on the summit ridge of Kangchenjunga. Then back to bed. Breakfast at 8:30 – a strictly British table with shredded wheat, bacon and eggs, toast, and Robertson's marmalade.

After breakfast, Norbu arrived and guided me to the small house that contained his family. Choni and his wife were awaiting me with Urgen, Norbu's younger brother. The girls were in school. After a convivial Tibetan midday meal I walked through the old town until it was time for afternoon tea with Mother Damien O'Donahoe. This was becoming a delightful ritual. Mother Damien gave me the girls' report cards that showed Yuden was making great progress and both girls' performance met with her approval.

The next day I flew back to Calcutta's Dum Dum airport and, with little delay, caught the Thai International flight to Kathmandu. Dr Kamal Shrestha, with whom I had worked in 1975, met me and took me to the Hotel Crystal in centre town that was to serve as the project's home-from-home for the next several years. Kamal had set up meetings for me with Dr Ratna Rana, Chairman of the National Committee for MAB, Dr Cherunjeevi Shrestha, MAB committee member, and various senior members of government and Tribuvan University departments. So I had a hectic and productive day.

On Kamal's advice, I rented a taxi the next day, and he led me on a highly instructive tour of the nearby Middle Mountains, including a reconnaissance of the Kakani ridge along the paved Trisuli Road to the lookout restaurant on its crest. In clear weather, I took many Hasselblad photographs. Kamal pointed out what he considered proof of the great threat of multiple landslides (strictly, debris flows). I carefully photographed one that had occurred at the height of the previous monsoon and marked the camera station for possible future reference (Fig. 41).[47]

On the third day, Kamal took me in different directions out from Kathmandu. The apparent instability of the very steep terraced slopes impressed me; landslide scars stood out in profusion, many cutting through flights of laboriously constructed agricultural terraces. I felt a twinge of guilt over my earlier statements in the Kathmandu 1975 MAB-6 report where I indicated scepticism about the claims that such damage was caused by deforestation. In retrospect, this quandary must have been influenced by the growing crescendo of news media alarm to the effect that the Himalaya were in a state of imminent environmental collapse. Nevertheless, it was still vital that real field data be acquired and studied so that any conclusions would be based on a rational analysis rather than emotional responses to assumptions. That approach was to be central to the entire UNU research under formulation.

On the final morning, Kamal accompanied me to the airport. His enthusiasm and determination for the UNU mountain field project to be centred on Nepal was somewhat overwhelming although later, as the Kakani project got under way, I was very thankful for his commitment. I reached Dehra Dun via New Delhi the same day.

Fig. 41 [top].   Photograph of a landslide (debris flow) that had occurred during the monsoon period prior to my 1978 visit. The full run-out extended an additional kilometre beyond the bottom left-hand corner of the view. I was told that one farmhouse was destroyed and two lives lost.

Fig. 42 [bottom].   Six years later little trace of the landslide remains, except for the appearance of the farm crops; it can be traced from the pattern of the maize crop. At the time the photograph was taken I was told that the productivity of the former landslide area was very close to that of the surrounding terraces.

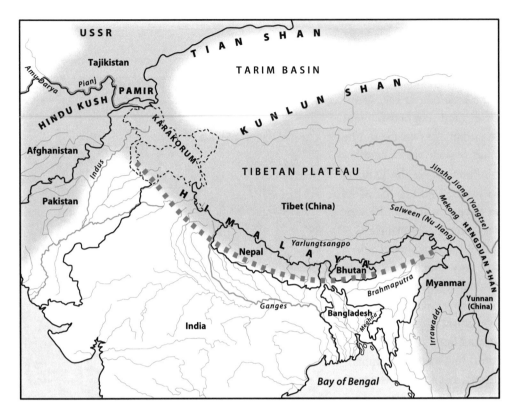

Map 3. The Himalayan and Tibetan Plateau region with the international borders and major rivers demarcated. The line of small squares delineates the "Himalayan Front".

In Dehra Dun, my host was Dr Jayanta Bandyopadhyay, and another new friendship was born that was to take us both all the way to UNCED and Rio de Janeiro in 1992, something incomprehensible at the time. He introduced me to the work of the Chipko Movement,[48] leading eventually to collaboration with its great leader, Sunderlal Bahaguna. Jayanta was a charming soft-spoken Indian, in contrast to the self-assertive Kamal. Within a few minutes of our first meeting, he presented me with a thick report co-authored with his wife, Vandana Shiva. I was up until the early hours reading. It was clearly indicative of the beginnings of forestry activism that led to collaboration with the Chipko Movement.

Jayanta arranged a series of very useful meetings for me with senior staff of the renowned Indian Forestry Research Institute. The second day, we made a long excursion by Land Rover into the hills, where I was introduced to a landslide problem comparable to that around Kathmandu and saw the environmental devastation of limestone quarrying on the steep ridge below the former Raj summer capital of Mussouri. Yet the problem of very limited access within Indian territory remained. Although the excursions to Darjeeling and Dehra Dun proved extremely valuable in broadening my Himalayan perspective, I decided to select Nepal as the main focus for the UNU mountain studies.[49]

The final stage in this long 1978 UNU exploratory mission was a three-day stop-over in Bern with Bruno and Beatrice Messerli. I explained in detail my initial proposals to

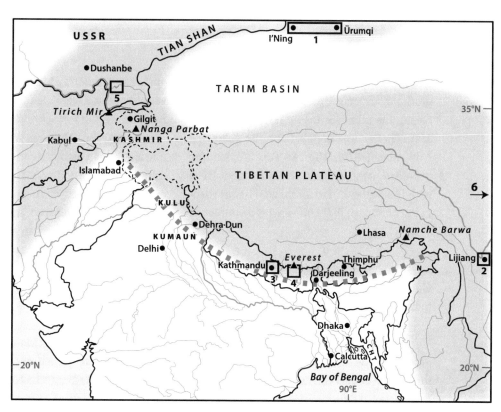

Map 4. The Himalayan and Tibetan Plateau region showing main cities and towns visited, prominent mountains, and major rivers. The line of small squares delineates the "Himalayan Front". The small open rectangles indicates the reconnaissance and intensive study areas: 1, Ürumqi – Tian Shan, NW China; 2, NW Yunnan – Lijiang, China; 3, Kakani – Kathmandu; 4, Khumbu – Everest ; 5, Pamir – Lake Sarez, Tajikistan; 6 (off map), Northern Hengduan Mtns., Sichuan, China.

Bruno. Eventually he agreed to participate in the venture as project co-coordinator on the assumption of Walther Manshard's concurrence. We discussed the advisability of a workshop in Chiang Mai that would allow us to invite a cross-section of researchers, both Thai and others, who could advance our planning for the proposed research amongst the hill tribes in northern Thailand (Chapter 6).

We formulated a threefold Nepal deforestation study coupled with mountain hazards mapping that would focus on the Middle Mountains, the High Himal, and the junction between the foothills and the plains. The mountain hazards mapping would also take in mapping of land use, study of agricultural practices through a range of elevations, and the training component that was stipulated

for all UNU projects.

Finally, I arrived home in Boulder, Colorado, exhausted but also exhilarated as two ambitious projects of some complexity and challenge appeared to be unfolding. Relaxation with family rapidly fused into preparations for six weeks of fieldwork in the Torngat Mountains of Northern Labrador. One of my family was able to take part. Tony at 17 was more than able to play an effective role and, to him, it was the adventure of a lifetime. This was the last piece of fieldwork stemming from my earliest post-doctoral research in Northern Labrador and Baffin Island, Canadian Arctic (Ives 2010, 2014).

Before departing for Labrador, I was able to exchange ideas with Gerardo Budowski and several other vital players and submit

two fairly detailed plans – one for a field programme, expected to last several years, in
Nepal, the other for the workshop in Chiang
Mai, Thailand, scheduled for the following
November (1978). The two sets of planning
unfolded concurrently and provided UNU
with the leadership to tackle vital issues within the context of highland-lowland interactive
systems.

Questioning the Causes of Deforestation in the Nepal Himalaya

During the 1978–1979 winter, plans for a
detailed study of the causes of mountain deforestation and its consequences were elaborated. It was tightly coupled with a scheme for
mapping mountain hazards as we considered
the two topics to be interdependent. Bruno
and Hans Kienholz became essential members of the team, and Bruno invited his Bern
University colleague, Professor Tj Peters, a soil
scientist. My INSTAAR colleague, geomorphologist Nel Caine, who was about to depart for
sabbatical leave in Tasmania, agreed to join
us in Nepal for three weeks. In addition to the
growing team of environmental scientists, it
was essential that we attracted a human geographer or anthropologist with experience in
the study of human response to natural hazards. The obvious choice was Professor Bob
Kates of Clark University, Massachusetts,
who was a leading scholar in this specializa-
tion and with whom I was acquainted. He declined my invitation because of over-commit-
ment but invited me to Clark for discussions
and introduced me to one of his completing
graduate students, Kirsten Johnson. I was
very impressed with Kirsten and offered her a
position on the forthcoming full-scale Nepal
reconnaissance. This was my first chance to
add a qualified woman to our team. Her commitment had another advantage: it would
ease the way for recruiting female Nepali
students from Tribhuvan University and further my general efforts to ensure mountain
field research opportunities for women. I was
convinced that it would greatly facilitate communication with local village women, something that male researchers would likely find
far more difficult.

I left home in early March for Bern where
Bruno and Hans joined me, and together we
proceeded to Kathmandu. Nel Caine and Kirsten Johnson arrived almost simultaneously.
Kamal Shrestha had taken care of all the logistical arrangements. According to protocol,
several Nepali officials were visited, including Ratna Rana. Bruno was able to establish
a useful liaison with Andreas Schild, who at
that time was director of the Nepal office of
the Swiss Association for Technical assistance
(SATA),[50] the bilateral agency commissioned
to assist with development in Nepal. I made
a good contact with staff of the Nepal–Australia Forestry Project. Both contacts proved
highly beneficial for the entire duration of our
project.

For three days, we travelled the routes I
had taken with Kamal the year before; there
was unanimous agreement that our intensive
Middle Mountain field research site should
include a large section of the Kakani ridge.
We also arranged a small group meeting with
several Nepali government and university

personnel as well as senior staff from the bilateral aid agencies, including SATA and the Nepal-Australia Forestry Project. We set up such a forum to explain who we were, outline our UNU commission, make a first sketch of our reconnaissance plans, and, above all, to ask for support and feedback.

Prior to our arrival, I had written to the chairman of Kathmandu's Tribhuvan University geography department to explain that, within the context of the planned research, there were three UNU overseas fellowships available for Nepali students. Would he assist us by working with the chairman of the geology department to advertise them and to set up a review panel that would include himself, his geology colleague, Bruno, Hans, and myself? This was done, and we devoted an afternoon to interviewing the ten applicants the notice had attracted. Kamal, as expected, had applied and with his German university doctorate in chemistry was the strongest of the group. He had shown himself to be highly effective in handling our essential local arrangements. We then unanimously selected Sumitra Manandhar (the only woman applicant) and Pradeep Mool. I asked that our selection be kept strictly confidential until I could ascertain confirmations from our three top-listed candidates.

Within two hours of concluding interviews and deliberations, our entire team attended a special reception organized for us by the U.S. embassy and the Kathmandu office of USAID. There we met numerous Nepali officials and senior international agency personnel – it was a valuable opportunity. The imperious Ratna Rana entered the room and immediately took me aside. He astounded me by indicating that I had a lot to learn about Nepali protocol and pointed out that the afternoon's selection process was flawed. Mool must be rejected and replaced by Tamrakar. When I asked him how he came to know the results of a confidential interview process, he casually explained that was another indication that I had a lot to learn. I asked for time to think through his advice (instruction!).

As I turned away from him, I spotted Corneille Jest in the act of shaking the hand of the U.S. Ambassador. I quickly approached him for advice. He began by joking that, if I had asked him to serve on the interview panel, none of this would have occurred. He next offered the advice that, because Ratna Rana had the power to cause our project considerable difficulties, I should accept Tamrakar (who happened to be a family relation of Ratna Rana) and try to find another way to accommodate Mool. As I had previously negotiated with Ratna Rana that the selection process would be on merit alone, I was rather upset, but appreciated the advice that Corneille, who had had extensive experience in these eventualities, had offered. I returned to Ratna Rana and agreed to offer a fellowship to Tamrakar, but also explained that I thought the situation could be resolved by a request to UNU to fund a fourth fellowship. In that case, I would prefer to let the initial offer to Mool stand. I then had to smile politely as Ratna Rana attempted an additional put-down by congratulating me on my "fast learning process". I was so angry over what had happened that I had determined, if necessary, to use my own funds to cover Mool's fellowship.[51]

The following day was spent in additional political and courtesy visits and final preparation for what would be the first real mountain venture of the UNU project for all of us – reconnaissance of the Khumbu area that would put us in the close vicinity of Mount Everest (Sagarmatha).

We assembled at the airport after an early breakfast. Our mission now depended on a successful RNAC Twin Otter flight to Lukla, a mountain valley airstrip established by Sir Edmund Hillary. At this season (March–April), there was a danger that local valley winds would develop during the mornings and that the increasing turbulence would cause cancellation of flights. There were four flights scheduled per day, even at that early period in the development of mass trekking to the Mount Everest base camp. Often, one or more had to be cancelled. It was important to obtain control of the first flight and our group, now augmented by Pradeep Mool and Rabindra Tamrakar, amounted to almost a full complement of the Twin Otter. This is where Kamal's rugged determination proved invaluable, even if embarrassing.

The first flight was fully loaded with mountain trekkers who had arrived at the airport before us, taking even Kamal by surprise. To my astonishment, Kamal boarded the plane as if he were an airport official and announced that the passengers must deplane and take an alternate machine that would appear momentarily. Once the plane had been vacated, Kamal urged us all aboard. All of this had happened so quickly that it took several minutes for us to realize the extent of Kamal's infamy. We were already half-loaded when comprehension hit. There was a comment that we couldn't allow this to happen as it was not our turn. The best I could do was to urge, following Corneille's advice under different circumstances: "When in Rome . . . ".

Soon we were in the air and enjoying a flight along the mountain front. Descending into Lukla was even more impressive – perpendicular to the trend of the valley and up the steepest landing strip I had ever seen.

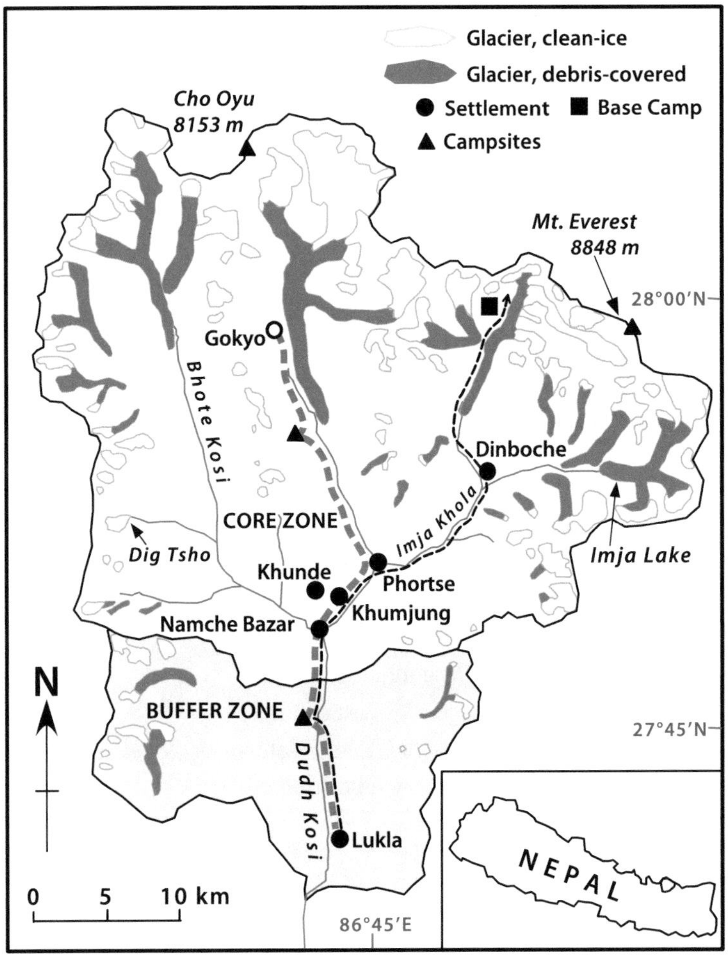

Map 5.   Sketch map of the Khumbu Himal. The dotted line indicates the route followed by the 1979 reconnaissance mission to Gokyo via Namche Bazar. The main trekking route to the Mt Everest base camp is also shown together with Imja Lake and Dig Tsho.

After landing we were relieved when, little more than an hour later, the second flight touched down, carrying the group that Kamal had outrageously ousted from the first flight.

The Gateway to Everest

Now we had reached the "gateway to Mount Everest." Sagarmatha (Mount Everest) National Park had been gazetted only three years previously, and it would take us a day-and-a-half of walking to pass through its southern entrance and reach park headquarters in Namche Bazar. The next event was a meeting with our Sherpas. Another great surprise – aside from our Sirdar, cook, and assistant cook, we were greeted by twenty young women in traditional Sherpani attire, giggling and waving their arms with enthusiasm.

It was a sunny, cool morning and, at this relatively low altitude, ideal for hiking. However, it was quite early in the year so the rhododendrons had barely begun to blossom. Our porters took on their backs all the heavy gear, including our rucksacks and pack-frames, tents, food, fuel, cooking utensils, and set off at a brisk pace laughing, singing, and playing tricks on each other. We walked up a well-trodden trail along the Dudh Kosi (river) about halfway to Namche, expecting that soon several 8000-metre peaks would be in view. There were very few other people on the trail, and we enjoyed a good day's hike arriving at our pre-selected camp site shortly after 4 p.m. to find that the Sherpanis had already set up the tents and hot tea was awaiting us.

On the second day, we arrived at Namche

Fig. 43.   Namche Bazar as it appeared in March, 1979. While the internationally-renowned trail to the Mount Everest base camp was already luring Western trekkers, their numbers were still modest and we passed only a few tourists.

Bazar. Before we had reached the confluence of the Imja Khola and the Bhote Kosi (which together form the Dudh Kosi) and the final steep ascent to Namche, we noted what appeared to be recent gravel terraces on either side of the main valley. They indicated that there may have been a considerable flood in the recent past, which had deposited extensive gravel fill, subsequently entrenched by the Dudh Kosi, leaving the still partially unstable terraces on either side. It was our first view of the effects of a *jökulhlaup*, or GLOF

Fig. 44. Mount Everest rears its head above the Nuptse ridge as high clouds pour eastward from its summit pyramid and from that of Lhotse. Twilight engulfs the lower Imja Valley (1979).

(glacial lake outburst flood) as they were to become known. We learned later that this particular event had originated beneath the slopes of Ama Dablam and was first studied by the use of remote sensing techniques (Buchroithner et al. 1982).

The day was warm and sunny. I had made the mistake of giving my anorak and woollen pullover to add to the load of our Sherpanis. In our enthusiasm to photograph everything, we were separated from them. We toiled, perspiring, up the steep slope towards Namche, where we planned to spend the next two nights, partly to acclimatize to the altitude. As we gained height more and more of the high mountain summits came into view. Soon we were able to see far to the west up the valley of

the Bhote Kosi and towards the Trashi Laptsa pass that leads into the Rolwaling Himal.

Hans and I reached Namche together as the sun was passing out of sight behind the western mountain rim. Nel and Bruno had been ahead of us although we had lost sight of them. The others were quite far behind. Now the sun had dipped behind the western summits and, without protective clothing, we were quickly becoming chilled. Not knowing the location of the lodge where we were to stay the night, we were reduced to slapping our arms across our chests as we continued to lose body heat in the relatively thin air at 3400 metres. By the time Kamal and our Sirdar reached us, I was beginning to feel the effects of altitude sickness. Fortunately, our lodge was quickly located and my Swiss colleagues, who were carrying a virtual Swiss Army pharmacy with them, were able to supply the much needed antidotes. In later years, this obvious lesson would have to be learned the hard way by hundreds of trekkers.

I rested at the lodge most of the next day. The others walked up to Khunde and Khumjung and visited the national park headquarters. By mid-afternoon I was sufficiently recovered to walk up the trail above Namche and photograph a tranquil sunset on Mount Everest with the Thyangboche monastery almost lost in the lower valley twilight.

On our fourth morning in the Khumbu, we set out for Gokyo, a moraine-dammed lake on the margin of the Ngojumba Glacier below Cho Oyu (first climbed by our colleague Helmut Heuberger in 1953, an 8000-metre triumph almost forgotten because of the ultimate feat of Hillary and Tensing on Everest the same year).

The decision to aim for Gokyo had been reached after considerable discussion. Ratna Rana had baited me about our plans to go to the Khumbu in the first place – he had described it as an example of distorting our UNU mission to walk up to the Everest base camp as a group of tourists and at public expense! This was one of the reasons behind the decision to begin our actual fieldwork on the Kakani ridge, rather than in the Khumbu.[52]

In perfect weather and after three kilometres, we reached a point just east of Khumjung where the trail divided. The right-hand fork descended to the Imja Khola crossing, en route to Thyangboche and the Mount Everest base camp. Our way continued high along the

Fig. 45.   The main Everest trekking trail above Namche. The steep slope appears dry and devoid of higher plants except for the occasional mature conifer. The multiple terraces have been produced by the trampling of generations of yaks. This type of landscape was described as the result of recent indiscriminate logging.

hillside almost due north until we obtained a fine view of the traditional Sherpa settlement of Phortse, perched on a terrace remnant on the far side of the valley. This was an impressive photo stop, including an excellent view of Phortse itself with Ama Dablam and other peaks in the background – the cover photo for *Mountain Research and Development* for the next ten years – and it also became a critical discussion point.

The view downstream, with the Imja

Fig. 46.   We pause for intensive discussions where the main trail descends to the Imja Khola. The snow-bound village of Phortse sits amidst its stone-banked fields on the remarkable terrace remnant. Ama Dablam dominates the right-hand horizon (1979).

Fig. 47.   We are close to our highest camp near the Sherpa summer settlement of Machhermo and the terminus of the Ngojumba Glacier. The small herd of yaks has been moved up-valley even though the high-altitude grazing season has barely begun and pasturage is by no means abundant.

Khola far below, displayed long barren slopes with little vegetation and cut by innumerable rough and short stretches of terrace (the product of trampling by yaks). On the opposite side of the valley, with northern exposures, dense coniferous forest was a conspicuous feature. One of the factors that had induced the Government of Nepal to gazette the Sagarmatha (Mt Everest) National Park in 1976 was the ardent claim by forestry advisors that, unless the protection of national park status was effected very quickly, extensive deforestation leading to environmental catastrophe would likely follow. This had been emphasized by the writings of Professor C. Fürer-Heimendorf, a renowned anthropologist, and others to the effect that entire mountain sides that had been in dense forest in 1957 had been stripped bare by 1977.

Challenging Conventional Wisdom

Among the scientists, we agreed that emotion and politics may have coloured the environmentalist claims of imminent disaster. Here lay an important stepping stone for what became one of the UNU project's main challenges (if not the primary one). The field evidence was beginning to augment our doubts about the simplistic explanations that had become conventional wisdom. Another talking point was the rather isolated situation of Phortse. The tangible benefits of trekking tourism and mountaineering were already becoming evident along the main trails throughout the Khumbu – rapid growth in the number of tea houses and lodges and the hire of yaks for transport, porters, and Sherpas for high-altitude mountaineering support. Phortse, because of its isolation, was clearly being overlooked. A similar situation prevailed in the upper reaches of the Bhote Kosi that remained closed to tourists at that time.

After lunch, we maintained a steady pace, this time with our baggage train in close contact. We eventually reached the small summer grazing settlement of Machhermo and the valley leading to the Machhermo Glacier. Within a few hundred metres we came upon a pleasant camp site. Tea was prepared by our assistant cook while we watched the Sherpani erect a village of tents under the direction of the Sirdar, and we settled down to a relaxed discussion of the day's experience. The campsite, at an altitude of about 4500 m, afforded a fine view of lesser Himalayan peaks with numerous small glaciers. Kyajo Ri at 6189 m dominated the valley while to the east Taboche (6367 m) and Jobo Lhaptshan (6440 m) formed the eastern skyline.

Morning dawned bright and clear. We planned to make this our final day of walking northward and after breakfast we set off towards the Ngojumba Glacier, leaving our Sherpani support crew behind to spend a leisurely day awaiting our return. A well-marked trail followed the depression between the lateral moraine of the glacier and the mountainside and we passed two small moraine-dammed lakes. Eventually we reached a much larger lake and the summer grazing area of Gokyo that has since acquired tea houses and a small lodge to become a minor trekking tourist destination. At about 4750 m, a moderate height, several of us were feeling mild effects of altitude. The attraction of Gokyo is primarily the neighbouring mountain top, little more than a steep walk of about 600 m that affords a view of the upper section of Mount Everest and several of its neighbouring peaks to the southeast. The majority of us were content to climb to the crest of the lateral moraine, a suitable lunch spot with an extensive panorama of the glacier with its heavy cover of debris intermingled with the occasional small melt pond.

After lunch, we retraced our steps of the morning, assembled our helpers with their pack-loads, and continued southwards to the point where the trail descends to the river bed, a little upstream of Phortse. Here we chose an excellent camp site, especially as the weather was closing in, and enjoyed a convivial warm evening, despite the first hints of snow as we eventually crawled into our sleeping bags.

A Very Human Cross-Cultural Diversion

As I poked my head out of the tent the following morning, I was dazzled by sunshine and snow. I took a short walk to enjoy the completely transformed landscape and accidentally ran into a raucous snowball fight amongst our helpers. Our assistant cook, a cheeky young Sherpa about 18 years old, had surprised two of the Sherpanis who were in the process of morning ablutions. For most boys that I can remember, the presence of snow and girls is an irresistible temptation and the two Sherpanis were sitting targets. But they began to return fire and, furthermore, soon had two reinforcements. At this

point, the assailant received back-up from the cook. Eventually most of the girls entered the fray causing the cook to flee and the mischievous boy to be physically overcome. Amidst much squealing and laughter, he was hoisted off his feet and his trousers were pulled down and stuffed with snow. He was then lifted off the ground by his belt and arms so that his trousers were firmly back in their proper place. Once the cold wet snow had completed its task, the girls collapsed in gales of laughter, the moral to this tale being that, if you are determined to throw snowballs at a group of Sherpani girls, make sure that your own are better protected.

After a hearty breakfast we were on our way, this time through ankle deep snow. We reached Namche after a very gentle walk and had several hours to wander about this interesting settlement in the early stages of what was to become an extensive transformation ("development") that has continued to accelerate until the present. The rapid growth of Namche, including new houses, hotels, lodges, and tea houses was assumed to have a serious deleterious effect on the local environment, especially due to consumption of local timber.

We spent the following day in Namche. It was Saturday, market day, and the focus of a remarkable mix of peoples, many of whom had walked up the trail all the way from the terai. The next afternoon we arrived at Lukla and settled in to comfortable lodge quarters close to the airstrip so that we were able to dispense with tents. The Sherpani made it clear that they expected a final goodbye party and we all worked to ensure that they were given a

Fig. 48. Lunch on the lateral moraine of the Ngojumba Glacier with Cho Oyu on the skyline. The glacier surface is covered with a deep mantle of debris, ranging from sand and gravel to house-sized blocks. The group includes (from right) Bruno Messerli, Rabindra Tamrakar, and our mischievous assistant cook (1979).

already married. This was hardly a reason for the Sherpani as surely he would find that two wives could be better than one. At this point, Hans sought shelter by introducing his inamorata to Bruno and me, and while she was momentarily distracted, he escaped into the darkness. For some reason that I won't go into, Bruno and I proved to be immune to amorous advances.

The last morning in the Khumbu dawned and we were away by 10:30 a.m. to the shouts and waves from our friendly team of helpers. In Kathmandu, there was time for relaxed discussion and for devising an outline for the research that we planned to undertake over the next several years. We were able to get together with our Nepali officials and aid agency staff to explain our findings and to elicit their response. In this way, we achieved for UNU the required formal commendation and encouragement for our proposed activities.

A leisurely breakfast on the attractive hotel roof garden of the Hotel Crystal the next morning became the place for taking leave as, in ones and twos, the team departed for the airport for flights in many directions. I spent an extra day in Kathmandu as I felt it wise to personally deliver reports of our future plans to Ratna Rana, Cherunjeevi Shrestha, and several other officials. I also needed time for discussions with our four Nepali team members. There were sufficient UNU funds, Pradeep excepted, for them to spend eight weeks the following July–August at the mountain research station in the Colorado Front Range. There they would join the three Chiang Mai students who had been selected by Gerardo Budowski and confirmed during my visit to

well-earned celebration. Food was abundant and was lubricated by ample amounts of rakshi, a traditional drink fermented from millet. Then there was the traditional Sherpa line dancing from which emerged the especially pretty Sherpani whom we had all noticed. Apparently, and much to his embarrassment, she had set her eyes on Hans. There followed several successive proposals of marriage. First Hans thought the explanation that such was impossible because he was leaving for Switzerland in two days would suffice. The would-be bride confessed that she would love to go with him to Switzerland. But alas, he was

Chiang Mai the previous February. This would provide an opportunity for them to mingle with the U.S. National Science Foundation high school students and undergraduates, together with our own and visiting faculty and graduate students who would be based at the research station. This experience would give them a solid introduction to a wide range of mountain research and related field methods. My own route home would prove somewhat unconventional – overland to Darjeeling – a decision that led me to the brink of disaster.

From Kathmandu to Darjeeling, 1979
A Journey to Remember

It was mid-April 1979. I had just come down from the Khumbu with the UNU mountain hazards mapping team (see previous section). I wanted to go to Darjeeling to visit Choni and Yuden, and family, the Tibetan refugees whom I had befriended several years earlier (Chapter 4). The standard route from Kathmandu involved a flight to Calcutta (Kolkata), which missed by an hour the flight northward to Bagdogra-Siliguri, necessitating a night in Calcutta. This I was anxious to avoid.

Kamal Shrestha, my Nepali colleague, suggested that I take the RNAC flight to Biratnagar, located on the terai in eastern Nepal. There I could hire a Land Rover to take me the

rest of the way. "Kamal," I said, "are you sure that the route does not cross restricted Indian military territory in Sikkim?" "Absolutely no problem," replied Kamal, "do you think I would make such a suggestion to my good friend? And if there is no Land Rover, you can go into Biratnagar and catch a good express bus that will take you to the frontier, and on the Indian side you will get a Land Rover."

So here began yet another journey, but this time it was based more on optimism, wishful thinking, and misplaced trust, than

Fig. 49. A surprise attack with snowballs is quickly met by a far more vigorous response than the mischievous perpetrator had anticipated; she had copious reinforcements.

on careful preparation. Karkavita, the Nepali frontier checkpoint was the intermediate target, and I had my correspondence with the Indian Embassy in Washington, DC, that would explain who I was, together with papers from UNU describing my mission.[53]

At 8:30 a.m. promptly, the RNAC Twin Otter kicked up dust from the Biratnagar airstrip, already warm compared to our last days in the Khumbu. Hasselblad camera bag, briefcase, and heavy suit case were set in front of the new airport; beyond, a gravel strip, then dense jungle; a line of bicycle rickshaws neatly drawn up awaiting hire. No Land Rovers, so obviously, it was to be the "express" bus.

The journey into the bus station was not at all promising; somehow the suitcase, Hasselblad, briefcase, and Jack managed to perch on the rear of the flimsy bicycle rickshaw as we teetered into Biratnagar. Bus station? Seemingly, a hole in the ground. One dilapidated bus with a somewhat expectant driver waiting sullenly: "Yes, this is the express bus to Karkavita. But please, sahib, take the front seat now, besides the engine – there you will be more comfortable."

By now – mid-morning – it was decidedly warm and I was beginning to perspire. I sat in the bus as advised by the driver and noticed that it was filling remarkably quickly. The passengers, including pigs, tethered chickens, and a shifty looking goat, soon had occupied all the available space and were beginning to overflow onto the roof, along with my heavy case. I clutched the Hasselblad and briefcase closely. Within half an hour we were off.

The express bus was destined to stop every two or three kilometres. It seemed that at each stop some five passengers alighted and at least half-a-dozen boarded, until the roof was overflowing and some of the more agile were hanging from door and windows. I now understood why the driver had urged me into the narrower front seat, despite the proximity of the engine. No body, human or animal, could possibly squeeze besides or over me.

By midday, the temperature had nudged from warm to hot; that is, the outside air temperature. Inside the bus, the situation was somewhat different. The engine cooling system began to boil over. Eventually the fan belt parted. Remarkably, the driver had a spare fan belt, but his only tool was a crowbar. Enterprise, brute strength, and good luck do sometimes succeed, as they did on this occasion.

About an hour later we had a flat tire. No jack! So everyone out to roll back a series of logs from the edge of the forest! The back end of the bus was progressively raised by main force as, one by one, logs were slipped beneath it. The spare wheel, threadbare, was installed, and we were off again.

Just before 4 o'clock we reached Karkavita. The actual Nepali frontier post was about 500 metres beyond the bus stop, so I hired a village boy to carry the suitcase, an improbable item in this totally rural environment.

The Nepali border guard asked for my passport, which was a simple task, and my military visa, which was not, since I didn't have one: "Sahib must go to Kathmandu and take the airplane to Calcutta if you want to go to Darjeeling." I could not persuade our border guard to stamp my passport and let me through. What to do? There was no question of ignominious retreat, and the prospect of a night in the village was decidedly uninviting. So I thanked the border guard, left his cement block house, and promised my boy ten rupees to race across the frontier with my suitcase.

Now I was in India: not only in India, but in the restricted military zone of Sikkim. I was soon picked up by an Indian border patrol and taken to their headquarters. An interview followed with the officer-in-charge – pukkha English: an army officer's course at Sandhurst, England. I produced my correspondence with the Indian Embassy, displayed my UN credentials, and imperceptibly (I hope) accentuated my English accent. "Jolly good show, old chap," he commented, as I would have expected from a Sandhurst-trained Indian officer, but he went on, "I think we can spare putting you in jail if you will write out on these forms a full explanation – let's say, confession – and sign it." Which I did. He remarked admiringly on my lucid explanation, even congratulated me on my clear handwriting, and said I should be on my way. I indicated that my passport required stamping. "Sorry, old chap, but I cannot possibly help you there," he replied. "But," said I, "how can I enter India without having my passport stamped?" "You already are in India, old chap, but I wouldn't worry. You made such an excellent impact on me and persuaded me to let you on your way, I am sure you will be able to exercise the same successful persuasion on my colleague at the other end. Besides, old chap, you cannot go back to Nepal because you cannot persuade me to give you an exit stamp to let you out of India." After my third cup of tea (I am not a tea drinker), I shook hands, said greetings to the Queen-Empress,

and staggered, overloaded, onto the road.

Now comes the bad part! As I hit the narrow surfaced road, it was quite empty, and it was already late afternoon. The sun was a little way above the treetops; apart from the military camp, there had been no sign of human habitation. There was a strange feeling in the pit of my stomach – Land Rover? And, as if by magic, a Land Rover appeared, seemingly out of nowhere. It wheezed and coughed its way along the road toward me; stopped with a shudder. Three piratical characters emerged. The driver spoke very broken English: "Sahib, where you go? I take you for small sum." I had to admit, had I met these three in the Khumbu a week earlier, I would have thought of them as colourful, jolly fellows, as trustworthy as a full moon. Now I wasn't so sure. The driver wore a Tibetan cap with earflaps, a bedraggled tunic, cotton pants, and a sash which served to hold a large kukri. His face was dark and gnarled and had a deep scar across the left cheek. His companions were equally dishevelled, apart from the scar and hat, although one of them appeared to have a deformed left arm. They were dirty and I could smell them at ten paces.

So we talked about driving to Darjeeling, and I offered them R150 for the ride. They accepted, loaded my case in the rear (short-wheel base) where two of them also sat. I pulled in beside the driver, dragging Hasselblad and briefcase after me.

On the third attempt, the engine fired, and we lurched off toward the east. About three kilometres on we teetered off the road and entered a jungle track: "Where are we going?" "A short cut," the driver replied.

A Close Encounter and a Time for Quick Thinking

It was getting late; the sun was into the treetops, and the treetops almost touched each other, making a tunnel of our trail with occasional open sections, as if the roof had fallen in. For the first time since I got on the Karkavita bus, I realized that my free outpouring of perspiration had changed to cold sweat; my chief sensation was fear, mixed with a sense of foolishness. A short distance farther, the engine spluttered and died: "Sorry, sahib, trouble with engine. Please get out." So out I got and walked about fifteen paces away from the vehicle. The man with the crooked left arm followed and stood besides me. The driver lifted the hood and appeared to tinker. The third man seemed to hesitate near the Land Rover. I felt I was taking part in a strange play where all the cues were missed: their hesitation, my fears, mounting. Then the driver broke the tension: "OK, we go now." We reoccupied our original positions and rattled off into the twilight.

By now I knew that I was in real trouble. I needed to think the situation through carefully and quickly, very quickly. Surely we would stop again soon, and this time the outcome could be much more uncomfortable. My father's advice to me as a boy came rushing back, since I suspected I was in for a fight – I would certainly put up all the resistance I could. So, as Dad used to say: When about to be attacked by superior numbers, you must strike first and hit the nearest one as hard as you can, preferably on the nose to produce a lot of blood. But the young thugs that sometimes frequented the Grimsby fish docks in my youth were a far cry from the three characters armed with kukris, even though they were much smaller and lighter than me.

The inevitable stop came five or ten minutes later: "Sorry sahib; engine trouble again; please get out." The performance of the first stop was repeated. I walked about a dozen paces and turned to face the Land Rover as the driver lifted the hood. The third man didn't hesitate this time and began to walk toward me. The second man was already standing on my right side. I tensed, realizing that I couldn't afford the barest slip. I managed to move quickly and surely. It seemed that in a single flow of my body I had his neck in the crook of my right arm, had lifted him off his feet onto my hip, and had pulled his kukri out and stuck it into his ribs with my left hand.

Again it felt like a bizarre scene from a play, but this time the cues were there and were met with precise response. My uplifted victim hung limp, barely moving. The other two screamed together, the driver in English, pleading that I had made a mistake, that they were my friends. I jiggled the tip of the kukri into flesh, far enough to draw blood and produce a most convincing scream, tightening my hold on his neck. I told the other two to throw down their kukris or else I would kill their colleague. A further protest, another press of the kukri, another cry of anguish, and the other two kukris were on the ground. I told them both to get into the front seat. I carried my victim to the passenger door, his tunic now clearly showing that I had cut him, pushed him in. Then I scooped up the two kukris and climbed into the back behind them. I told the driver to drive to Siliguri, one false move and they would feel the knife.

We drove off into the fading light; within ten to fifteen minutes it was quite dark. My racing pulse kept up a heavy throb and an appalling feeling came over me as I realized how close I had been to killing a human being.

Sometime later we entered Siliguri. The main road, like any Indian town, was teaming with people, on foot, on bike, with bullock cart, standing in groups, and stray cows wandered casually across our path. All was dark and dusky beneath a few scattered lights. As we felt our way toward the centre of town I became anxious again. What next? How was I to get out of this? We came upon an illuminated taxi stand with a young man, white shirt, tie, sitting on a bench. I told the driver to stop. He did. I asked the young man if he could take me to Darjeeling by taxi. He sprang up with vigour: "Don't worry sahib, I will have my father call the police." "Please, no," I urged. "Could you lend me R150 so I can pay them off?" "Of course, but you don't need to pay them; they are bandits." Nevertheless, they were paid and quickly disappeared into the gloom.

"Dr. Dilip Dey is expecting me at the Gymkhana Club in Darjeeling. Can you drive me there?" "Oh, Dr Dey is a good friend of ours, of course I will take you." "Then he will repay the R150 and also cover the taxi fare because I have no Indian cash and crossed the border illegally." "You still shouldn't have paid them." "But I said I would." "Some of you English are very strange," he said. Of that I am oddly proud; nevertheless, it was probably the smartest thing to do and so have them disappear without a murmur. I was highly conscious of my illegal border-crossing and lack of a visa or a passport entry stamp so that I did not think I would be well placed if faced with the prospect of giving evidence to the local police.

We took the long twisting drive up two kilometres to Darjeeling under a starlit sky. I remarked on my new driver's erudition. He told me that he had completed an MA in geography at the University of Illinois but couldn't get a professional job when he returned to India, so had gone back to working with his father's taxi business.

We reached the Gymkhana Club close to midnight. Dilip, his face a purple hue, was waiting in a state of acute anxiety. "Oh Jack, when you didn't arrive from the afternoon plane, I became afraid that you had tried to come overland. Do you know that several travellers have been killed, and others robbed and stripped and left in the jungle in recent months? You never should have tried that route." "Dilip, in 1975 you told me it was the route you had used to come to our meeting in Kathmandu; I merely tried it in the opposite direction." Two more gin and tonics, and we both fell exhausted into our beds.

The following day, I had the opportunity for another visit to Choni's family and to Loreto College. In the afternoon, I entered the college for my teatime appointment with Mother Damien. She sat me down in her attractive study, and we took the occasion to look out once more onto the great mass of Kangchenjunga, this time almost entirely lost in great banks of cloud. We discussed Yuden's progress, which Mother Damien thought was excellent; she questioned me about my work with UNU and then called in a teaching nun to give me yet another tour of the college and to enjoy this tranquil sanctuary of female learning. And once more I was able to visit Yuden and Lhazom and be further encouraged by their progress.

I left the college and spent the remainder of the afternoon walking the streets of Darjeeling that never failed to fascinate me and eventually returned to the Gymkhana Club. After dinner with Dilip and several of his friends, I was invited to play billiards with one of the tea planters from "down the road." I astonished myself by giving him a fair game although there was never any likelihood that I would win. The next morning, I received a surprise package – two pounds of the world's finest tea – "for allowing me to beat you at billiards" written as part of an eloquent personal note by my billiards partner of the previous evening.

As Dilip drove me down to the airport, I explained carefully that my passport would show no Indian entry stamp and asked his advice. He warned me that I might be running into trouble. He was silent for some moments, deep in thought. Then he advised that, although I might not like doing so, I should pose as a very important Englishman, as if the Raj still existed, and berate the senior officer for his gross inefficiency for not having my passport stamped on arrival. In those days, Canadians did not require visas for India, although special visas or passports bearing a military stamp were necessary for entry and exit from the Himalayan military zone. My passport was also lacking a Nepalese exit stamp, but this was also overlooked.

However, Dilip's prophecy of portending trouble was correct. Although I felt very uncomfortable, I put on the act as advised and despised myself when I found how well it worked. But, unthinkingly, I had given the officer opportunity to square the pitch. When my hand luggage was being inspected a few minutes later, it ensued that I had been foolish enough to forget to take my Swiss Army knife out of my briefcase and transfer it to the checked luggage. My only recourse, when the baggage inspector called him over, was to berate him further: "So now you have a chance to cause me trouble. Let me see, my man, what you will do." He let me board the plane for Calcutta with knife in briefcase.

Early Thoughts on a Predicted Himalayan Disaster

Following the three separate visits to Nepal, I had become increasingly sceptical about the widely publicized scenario that the Himalaya were facing imminent environmental disaster. The prediction appeared more and more likely to be based upon a combination of emotion, out-of-date university textbook hypotheses, and sheer politics. I could not deny that the region was facing serious problems. Nevertheless, we had begun to question whether the poor mountain farmers were really the perpetrators of environmental degradation, as they were almost universally perceived. Were the farmers of the Middle Mountains as well as to the farmer-traders of the High Himal the culprits, or were they the victims of gross prejudice and speculation? From our still relatively limited contact, we had decided to approach them as intelligent equals and to seek their advice, rather than enter their territory as "educated Western experts" to set them aright.

The addition of socio-economic and anthropological enquiry was by no means sidetracking our original intentions of developing a system of mountain hazards mapping. It would provide the underpinning of the entire enterprise. Therefore, a large amount of basic data on a multidisciplinary level would be required. It would be useful to approach this through a series of questions. The first group of questions refer especially to the Kakani area that we were accepting as representative of the Nepal Middle Mountains:

1. What was the history of Himalayan deforestation, including the period long prior to the present (that is, 1950 to 1979)?
2. Could we determine the number of landslides that occur annually and study the relationship of landslide density to bedrock type, slope orientation, altitude, type of agricultural terrace, and crops grown?
3. How did the farmers respond to the landslide problem?
4. What was the timing of the landslide incidence in relation to the onset of the summer monsoon?

In all of this, we found ourselves in step with the thinking of the staff of the Nepal-Australia Forestry Project. Similar questions, with modification, would be applied to the Khumbu area. We should also search for landscape photographs taken on the early expeditions to climb Mt Everest, as they might provide a means of comparing degree of vegetation change since the early 1950s. Finally, the various statements of rapid deforestation made by some of the major agencies must be tested: for instance, the World Bank (1979) announcement that "at the present rate of deforestation" there would be no accessible forest remaining by the year 2000. How was that "present rate" determined and could it be checked for accuracy?

All the questions would be more relevant if we could learn more about the way of life of the hill farmers. The variety of land-use practices by the several different ethnic groups would cover a large range in altitude. The third test area that was originally proposed as part of the project, the contact zone between the mountains and the terai and Ganges floodplain, still needed to be reconnoitred so would have to be left in abeyance for the time being. We certainly had enough work on our hands.

Fig. 50.   An oblique view of the Potala, the highlight and shadow showing the magnificent proportions of this sacred Buddhist shrine and former home of the Dalai Lama. In 1980 extensive repairs were already underway in preparation for its conversion to a major international tourist attraction.

CHAPTER 8

The China Connection
Privileged Entry to a Closed Country

In early September 1978, I had received word from the U.S. Department of State that a team of senior Chinese academicians had been given permission to tour a number of research establishments in the United States. As a visit to INSTAAR's mountain research station had been one of the principal requests, could I serve as host? I agreed with enthusiasm. While I was pleasantly surprised, I was amazed that anyone in China had even heard of our mountain station because, in those days, China remained a closed country.[54]

I was told to expect a group of about eight academics with Professor Huang Pingwei as leader. He was director of the Geographical Institute of Academia Sinica (Chinese Academy of Sciences) and hoped that his group could spend three or four days with us, the main interest being to visit the mountain research station and, if weather permitted, our research sites above timberline on Niwot Ridge. As the date of their anticipated arrival was set for November, I could see that we had a challenge on our hands – there could be a problem with deep snow. However, Mischa and Olga Plam, residents at the research station, were ecstatic, especially as two of the

intending visitors had been Mischa's graduate students in Moscow all those many years ago before he had come into conflict with the KGB (Chapter 5).

The first full day of the visit was a relaxed tour of the University of Colorado campus in Boulder ending with a dinner in the faculty club, attended by the deans of Arts and Science and the Graduate School, together with the Dean for Foreign Relations. The next morning began early with a drive up Boulder Canyon, with snow on the ground just above Nederland at an altitude of about 2500 m. Next a tour of the research station and finally, the highlight of the visit, an over-snow journey through the subalpine forest belt and out onto the tundra of the ridge crest at about 3500 m. We mobilized all our over-snow vehicles and distributed parkas, goggles, headgear, and gloves to our visitors, who were dressed in dark business suits. I took Professor Huang Pingwei behind me on a skidoo. This resulted in a very amusing "accident." During a pause before the final steep climb to the ridge crest, my important passenger stepped off the machine and plunged almost to his armpits in deep soft snow. We extracted him amidst gales

J. D. Ives, *Sustainable Mountain Development*, https://doi.org/10.1007/978-3-030-96029-2_8

of laughter, his leadership decorum and somewhat formal personal image being softened in the struggle to pull him back onto the vehicle. It was then that his normal personal warmth and conviviality showed through, the beginnings of a long and significant friendship.

On the ridge crest, we briefly discussed the alpine environment. Roger Barry gave a succinct account of the alpine weather and climatology, and Mischa drew comparisons with his beloved mountain station in the Caucasus. The high cloud cover afforded us good views of the surrounding mountains, and I was in no doubt that several slide shows of a part of the Colorado Rockies would later be projected in Beijing.

Dinner at the research station was a great success. Olga had decorated the dining room with Chinese lanterns and a huge array of candles, and a big log fire in the ancient cast iron stove maintained a warm atmosphere. We had hired a cook, but Olga had added her outstanding talents to preparation of the meal. And there were toasts and speeches galore. Professor Huang Pingwei introduced us to *moutai*[55] and Mischa's addition of vodka made for a wonderful Russian-Chinese-American statement of mountain solidarity. Roger delighted the Chinese by addressing them in Russian, which most of them had learned during their graduate student experience in Moscow. Roger and I presented Professor Huang Pingwei with a copy of our 1000-page book *Arctic and Alpine Environments*. All in all, it was apparent that a high level of mountain rapport had been established. At the time, I had no idea that this rapport was already set to exert a great influence on my still-not-fully formed international mountain research schemes. Some of this came to the surface the following morning as we were taking our guests to Denver's Stapleton Airport.

As a departing gesture, Professor Huang Pingwei astounded me by asking if I would accept an invitation to visit Beijing and Lhasa with the possibility of travelling through the Himalaya to Kathmandu. After I accepted with profuse thanks, Professor Huang explained that Mr Deng Xiaoping's determination to open China to the West was to be heralded by invitations to an international group of mountain scholars to attend a conference in Beijing on "The Geology, Geological History, and Origin of Qinghai-Xizang Plateau." This was to be followed by an extensive excursion to Lhasa and vicinity, ending with an overland journey through the Himalaya to Kathmandu. With a friendly joke that alluded to my English background, Huang Pingwei said that it could well be the first time since the 1904 British imperial military "raid" by Younghusband[56] that such a formally approved journey had been undertaken. A final goodbye handshake led to a promise that he would see me in Beijing, probably in the early summer of 1980.

Journey to Lhasa, the 'Forbidden City', and Beyond

The anxiously anticipated invitation to attend the conference in Beijing arrived in the early days of 1980.[57] It was extended to 64 foreign scholars in all, together with more than 200 Chinese. The conference was planned for 25 May to 1 June 1980, to be followed by a flight to Chengdu and on to Lhasa restricted to about 80 of the participants with many of our Chinese counterparts serving as guides.

In the 1970s and 1980s, up to the time of receiving my treasured invitation from Beijing and arrival on Chinese soil, I had passed through a remarkable series of experiences. They included the highly successful reconnaissance of the Chiang Mai area of northern Thailand and the Nepal Middle Mountains and Khumbu Himal (Chapters 6 and 7), declaration of Niwot Ridge as a UNESCO Biosphere Reserve by U.S. President Jimmy Carter, and a colourful addition to the mountain research station summer school of international (Nepali and Thai) UNU fellows, an outcome of the training aspect of the UNU research programmes that were in the process of formulation. In addition, INSTAAR had acquired the transfer from USGS of the World Data Centre-A (Glaciology), of which Roger Barry was appointed director.

In the midst of these palpable successes, which had also included designation of the University of Colorado as an associate institute of UNU (with Harvard University the only other U.S. associate at that time), I resigned as director of INSTAAR. This enabled me to devote myself to what I had come to perceive as a significant opportunity that would take me far from my then existing responsibilities.[58]

First International Mountain Conference in China: Beijing and Lhasa

My arrival in Beijing on 23 May 1980, was intriguing. I was met by enthusiastic members of Academia Sinica and, after very cursory customs and immigration processing, was whisked into a waiting car and driven to a large barracks-like hotel – architecture of the Stalin era – somewhere in the centre of the city. There I was allowed to sleep undisturbed, a most welcome aspect of hospitality, considering the long air time from Colorado and a heavy dose of jet lag. Eventually, hunger prevailed and I found my way to the dining room to meet several old mountain friends, including Bruno Messerli and Corneille Jest, ensuring that our IGU and UNU partnerships would be further strengthened.

When most of the foreign participants had assembled, we organized an informal meeting and, by acclamation, elected as the doyen of our group Professor Ardito Desio, the well-known elderly glaciologist. In 1954, he had been leader of the first expedition (Italian) to reach the summit of K-2, the world's second highest mountain.

By mid-afternoon, we were all driven to the Great Hall of the People in Tiananmen Square for the official welcoming ceremony. I found it a great thrill to enter this politically very powerful building and certainly felt honoured as one of the few foreign scholars chosen to be the vanguard for the "opening" of China and the Qinghai-Xizang (Tibetan) Plateau. The foreigners assembled in a vast hall, along with more than two hundred of our hosts, to await the entry of Mr Deng Xiaoping.

Deng Xiaoping's address of welcome was electrifying and he made it clear that we were chosen because of our well-understood interest in the Tibetan Plateau and expertise in mountain problems in general. He explained that he hoped this would be the beginning of international research collaboration with China. He informed us that it was on his personal dictate that we would be the first group of guests to travel overland from Lhasa, through the Himalaya, to Kathmandu.

Professor Desio, more than 80 years old, very tiny and slim, carefully attired, stepped forward to respond on behalf of the entire group. After a short speech, he shook the hand of Mr Deng, who amazed us by standing back a pace and remarking, "Professor Desio, you are even shorter than me." To which Desio quickly responded, "No Sir, you are shorter than me." The Chinese leader immediately called two soldiers forward and instructed them to produce a measuring device. After a very brief exchange of light-hearted banter, the measuring tools were produced (I wondered at the time whether this had been set up previously, although Desio denied it) and, like small boys in competition, the two stood back-to-back and strained for maximum height. After careful consultation with the soldiers, Mr Deng delivered his compelling judgement with uttermost diplomacy: "Professor Desio, we are exactly the same height." Patting his belly, he went on: "But you are smaller than me." – to a burst of laughter across the hall.[59]

During the following celebratory dinner in the Great Hall of the People, I met Professor Huang Pingwei, who extolled his enthusiasm for the previous year's visit to our mountain research station. He then took me to meet Deng Xiaoping. On the way, he also introduced me to Dr Sun Honglie who was destined to work closely with me for many years. Mr Deng invited the three of us to join him for one of the many dinner courses. He delighted me by saying that he hoped I would be attracted to extend my mountain research interests to China. I was too politically naïve at the time to realize that he was issuing a serious invitation.

In Beijing, several days of paper sessions were interspersed with visits to the Great Wall, the Imperial City, the Ming Tombs, together with lavish entertainment that included the Peking Opera, where we certainly fascinated the regular local audience. However, our hotel gates were locked at 10 p.m. each night. We were told that this was to protect us from the never-ending stream of Chinese visitors who would otherwise be talking with us throughout the night. Already a rumour had begun to circulate that the real reason was that the authorities wanted to keep us strictly separate from local citizens. This point was belaboured by the American news media on our return home. I couldn't resist testing the supposition.

Early the third morning, I took out my Hasselblad camera with the large telephoto lens attached (so that it could be seen from a considerable distance) and went down into the courtyard well before breakfast. The massive iron gates were padlocked. As I approached, two soldiers saluted, leaped forward, and opened the gates, waving me through. I strolled about the neighbouring streets for the next 45 minutes taking photographs at will and having very affable, if unintelligible, conversations with local people on the streets. So much for rumour and propaganda.

The conference itself proved a remarkable experience. We learned a great deal, both from the presentations by our own group of foreign guests, and especially from the large number of Chinese contributions. The entire proceedings were published in English the following year by Academia Sinica Press in two hardback volumes totalling 3112 pages. In terms of assembly of papers, translation and editing, this was a triumph.[60]

From discussions with our Chinese colleagues, we learned that many of them had avoided the ravages of the Cultural Revolution as they had been part of a series of scientific expeditions to the Tibetan Plateau – the source of so many of the papers that had been presented during the conference. I also learned that my new acquaintance, Dr Sun Honglie, a high-altitude soil scientist, and Professor Shi Yafeng, who quickly became well known in the West as the "father" of Chinese glaciology, would be the scientific leaders of our excursion to Lhasa and the Plateau in the forthcoming days.

While talking with Huang Pingwei about the prospects of cooperation between the Chinese Academy of Sciences and UNU, I mentioned the university's approach to selecting scholars for visiting fellowships with associate institutions. He strongly urged me to talk with Sun Honglie. He also introduced me to his colleagues in the Institute of Geography, Drs Zhang Yongzu and Zhang Peiyan. I talked extensively with all three, who appeared very keen to spend a year with me in Boulder. Shortly thereafter, vice-rector Walther Manshard confirmed all three fellowships. Our new Chinese colleagues arrived in

Boulder the following September, thus laying the foundations for future collaboration.

We flew from Beijing to Chengdu on the morning of 2 June. Here we were to spend two nights before flying on to Lhasa. Our stay in Chengdu was a memorable experience. While we were advised not to wander off the main streets on our own because our hosts feared that our welcome might not be entirely friendly, we were met with a reception ranging from shy curiosity to enthusiasm.

I managed to slip away down narrow streets, trailed by a group of small boys who were fascinated by my red beard, freckles, and blue eyes, the likes of which undoubtedly they had never seen before. I stopped to make friendly gestures to groups of adults sitting on the steps of their houses. They welcomed me to sit with them and enjoy a bowl of noodle soup, laughing at my performance with chopsticks. They were even more enthusiastic when I showed them my camera and indicated that I would like to photograph the children. When I let one of them look though the Hasselblad telephoto lens, the ensuing reaction made me fear that I would never get the camera back again. I began to think that this kind of friendly reception was probably a worldwide aspect of humanity especially among poor people. It was a very humbling experience that I was to experience time and again over the following decades.

In the afternoon, Bruno and I took a walk through the large park in the centre of the city. It boasted an extensive lake with rowing boats for hire. A group of students approached and asked us to take a boat ride with them. They were learning English at the city university

Fig. 51.   This photograph was posed – the only way
to calm the close-encounter enthusiasm displayed by
almost all the Tibetans we met in Lhasa in 1980.

Fig. 52.   A surprise encounter that resulted in an accidental camera exposure. With the help of Corneille Jest, Tibetologist, I learned that the Tibetan man had lost his leg in the defence of Lhasa against the attacking Peoples Liberation Army (1980).

and were anxious to talk with us. Even Bruno's Swiss-English accent was treasured, but when I told them I was Canadian, I encountered cries of joy because their English language training tapes had been made in Canada. Four of the students rowed us out onto the lake and, once we were well away from the shore, they triumphantly announced that they had kidnapped us and our ransom was thirty minutes of conversation. I include these several acts of friendliness to indicate how one must always beware of official propaganda.

In a state of anticipation, the next morning we all awaited take-off for Lhasa (truly the Forbidden City, now about to become much less "forbidden"). The flight was remarkably smooth and afforded good views of snow-capped mountains, a sight that none of us foreign guests could have anticipated seeing only a year earlier. To my surprise, Lhasa airport was a gravel strip in the valley of the River Yarlung-Tsangpo and we had a long bus drive into the city. Corneille, that minutely precise observer, asked if I had noticed any difference in the general landscape between here and the Khumbu Himal, not so very far away – it was the almost total absence of Buddhist artefacts! Corneille ventured that half of the People's Liberation Army must have been employed in their removal.

On arrival in Lhasa, craning our necks to catch glimpses of the Potala, the former palace of the Dalai Lama, we were driven to what appeared to be a substantial military barracks. Here we unloaded and were addressed by our medical advisers. We were urgently requested to go to our rooms and lie down for at least three hours and under no circumstances to risk any physical activity until the next day on account of the altitude and our rapid ascent from the low altitude of Chengdu. We were also assured that we would have with us at all times abundant oxygen and other medical supplies.

It was on this occasion that I began to realize the deep concern of our hosts for our well-being. Not only the possibility of altitude sickness, but in general everything from headaches to the prospect of a sprained ankle were covered – we were very definitely special guests of Deng Xiaoping. Regardless of the good advice, one visitor was heedless and managed to slip out of the building undetected. Apparently, he was determined to run up the long flight of steps leading to the main entrance to the Potala. To our great embarrassment, we learned that he had collapsed halfway up and was recuperating in an oxygen tent. By the next morning, the medical prognosis was such that a special flight had to be made to get him back to near sea level. This was a most unfortunate beginning of our stay on the Plateau and prompted Professor Desio to make an official apology on behalf of all the visitors. It was most graciously accepted.

The next several days made us realize the extent of the privilege that had been afforded us. Visits to the Potala, the Summer Palace, the Jokhung Temple, and the Drepung Monastery made us feel that we were truly "on top of the world." However, Corneille took me aside and indicated that he assumed I would like to take portraits of Tibetan people and to talk with them privately: that meant, without our constant military escort. Corneille was fairly fluent in Tibetan and it was obvious that, with members of the People's Liberation Army at our elbow, there would not be the remotest prospect for any private contact. In fairness to our hosts, we sensed they were afraid that we could meet with some hostility and wanted to ensure that we were fully protected. Nevertheless, I followed Corneille out of his hotel room window and we walked, unhindered, into the Old Town. He led me on the traditional Buddhist walk around the Jokhung Temple within the flow of a large number of Tibetans, quite a few of whom were prostrating themselves for the entire circumambulation.

When the local people saw that we were alone, they flocked to embrace us. There was an impressive show of welcome. When we asked permission to take photographs, they were delighted. In fact, one lady holding a bunch of radishes, wanted to embrace me, camera and all. The only way I could get her into focus was to gently push her back and, in effect, pose her. Normally, I do not like doing that, but I did and so obtained one of my most beautiful portraits. I only regret being unable to send her a print, although she has appeared in several of my international photograph exhibits. In another instance, as we continued our circumambulation, we suddenly came across an old man and a small girl in tattered clothing. The man was seated, propping up a pair of crutches that had replaced his missing leg. My camera was at my waist and I immediately released the shutter, shooting blind. We talked with them and I learned through Corneille that the man had lost his leg defending Lhasa against the advance of the People's Liberation Army.[61]

The Potala and the Summer Palace were in the process of being renovated. We were offered what appeared to be unrestricted access. A high point for me was the quiet peacefulness of the Dalai Lama's study in the Summer Palace with many of his personal effects discretely set out, including his telescope which he used to focus on the crowds beyond his normal access. The virtual absence of monks in the Drepung Monastery left us with the impression that we had entered a long-abandoned ancient monument.

Our first long excursion beyond the limits of Lhasa was through the Nyantanglha Mountains to visit a site that was being developed to supply thermal power to the city. We were informed with what I thought startling frankness that China's hold on Tibet would depend heavily on solving the problem of providing heat and electricity to Lhasa. It was hoped that thermal power would become available within a short time. Was this the beginning of "eco-friendly" development?

Among the several surprises arising from our first excursion across the Plateau were the considerable stands of shrub forest at high elevation. Although they were restricted to less accessible and steep mountain slopes, at about 4800 m, they were well above what I had believed to be the Tibetan natural upper treeline. I suspected that the heavy use of timber over the centuries for construction of temples, monasteries, and palaces, as well as fuelwood, had significantly lowered the anthropogenic treeline below the "natural" one as portrayed in Western textbooks.[62] If only Carl Troll could have been with us!

Fig. 53.   On the roof of the Jokhung temple a solitary monk provides a human silhouette.

Fig. 54.   The Chinese search to harness thermal energy in the Nyantanglha mountains on the Plateau north of Lhasa. Note the dense patches of shrubby conifers on the lower slopes of the distant mountains (1980).

Fig. 55. The drive down the gorge through the Himalaya and so out across the verdant terraces of Nepal's Middle Mountains was impressive. We were told that more than a thousand lives were lost during construction of the highway.

Fig. 56. At the Friendship Bridge on the Nepalese/Tibetan frontier we are to part company with our congenial Chinese guides. Bruno Messerli, our tallest member, is standing next to me (in yellow waterproof) and Sun Honglie is on his right.

By Road to Kathmandu

After the last of the local excursions, we departed by bus on our long overland journey to Kathmandu. We retraced our inward route to the airport, crossed the Yarlung-Tsangpo River by ferry and then entered an extremely dry, mountainous region with beautiful salt lakes and tiny settlements, their emerald green field patches dependent on local irrigation from nearby glaciers. Several of the interior valleys displayed textbook examples of crescent-shaped (barchan) sand dunes. We passed remnants of the semaphore system that had been erected in 1904 by the Younghusband military incursion to ensure communications with the rear base in India.

The high point of the route, close to 5500 m above sea level, was the occasion for a photo-stop. To everyone's amazement, Professor Desio, always immaculately dressed in suit and hat, jumped out of the bus and began running up the mountain side. At his age, this was perceived as extremely dangerous and threw our Chinese medical advisors into a panic. They rushed after him, while others began laying out the oxygen apparatus. But they couldn't catch him! Eventually he paused and took a round of photographs from a point about two hundred metres above us. He then leisurely walked back to the bus and re-entered to a vigorous round of applause, looking surprised and clearly wondering why all the fuss.

We had a long stop in Xigatze to photograph the rather dilapidated but impressive palace of the Panchen Lama and the surrounding temples. This had been the site of a strong Tibetan resistance to the Younghusband military column, which had been overcome by the British deployment of field guns. We drove onwards to Tingri[63] where we had a distant view of Mount Everest (Chomolungma) from the very arid high plateau. We passed through several small villages whose inhabitants came out to meet us. They and their homes indicated extreme poverty.

Before leaving Lhasa, we had been warned that early monsoon weather had produced heavy rainfall and much snow at the higher elevations. As a result, avalanches and landslides had crossed the trans-Himalayan highway. However, this threat of disappointment was stilled when we reached Xigatze, where news reached us of Deng Xiaoping's personal involvement. He had ordered an entire division of the People's Liberation Army to clear the way – another impressive indication that he was the mastermind behind the enterprise. Certainly, the descent down numerous hairpin bends and between mounds of cleared landslide rubble was breathtaking.

Eventually we reached the Friendship Bridge, the border crossing into Nepal where our Chinese friends would leave us and return to Lhasa. We boarded a waiting bus on the Nepalese side of the bridge. Before crossing into Nepal, however, we took the occasion for a hearty celebration and many group photographs.

As we drove toward Kathmandu, we all sensed the dramatic change in the environment: from extremely dry and hypoxic plateau to the lush, moist, forested, and minutely terraced landscape, all within a few hours. Thus, we had completed a historic traverse.

From Kathmandu the group dispersed to their homes in all parts of the world.

One of the immediate outcomes of the journey to Tibet was that I received many invitations to give illustrated lectures. One at the Massachusetts Institute of Technology, arranged by my close mountain friend Frank Davidson, provoked an interesting incident. As on other occasions, the question period after my talk seemed interminable. Most of the interest seemed to be pointed questions with an assumed answer – that Chinese treatment of Tibetans, as well as their own poor people, was brutal. I was becoming tired of this, especially as it seemed to crowd out what should have been discussion of the potentially very positive aspects of future research collaboration. Finally I responded to one such question in a rather impatient and inappropriate manner; I supposed that, had I been a Tibetan and had been given a mystical choice to having my family overrun by India or China, I would have opted for China. Very unfortunately, the questioner identified himself as Indian. He was outraged and demanded an apology. For once, my wits stood me in good stead as I invited him to join me for tea so that I could answer at leisure. The chairman breathed a sigh of relief as my interrogator agreed and closed the meeting with no untoward reaction, except for a lot of student laughter. And over tea I made my peace and we parted on good terms.

This fascinating entry into a closed China, and especially to Tibet and through the Himalaya, laid the foundations for a long research collaboration that was to last well into the 1990s. Furthermore, it provided a vital element for the UNU mountain programme and was speedily followed by reconnaissance ventures into the mountains of northwestern and southwestern China.

Fig. 57.   Off-shoot of the main range of the Tian Shan: a high pass in the Dzhungarskiy Alatau range in Xinjiang Province, China, close to the Soviet frontier. The meadows are the product of centuries of transhumance grazing by Uigur and Kazak herders. They make an attractive pattern between the dense stands of evergreen trees (*Picea schrenkiana*) (1981).

CHAPTER 9

Beyond Lhasa: Opening China to Collaborative Mountain Research

The proposed collaboration with Academia Sinica proceeded apace following the unprecedented journey in 1980 from Beijing to Kathmandu via Lhasa. Dr Sun Honglie was offered a post-doctoral fellowship by UNU to spend a year with me as a guest of the University of Colorado, and I welcomed him to Boulder in September 1980. He was accompanied by the other two senior Chinese geographers whom I had met during the Beijing conference, Dr Zhang Yongzu and Dr Zhang Peiyuan, also UNU fellows.

On Professor Huang Pingwei's urging, I drafted a recommendation for joint Academia Sinica[64]–UNU field research. I diverged from our original thoughts and proposed instead a less ambitious beginning because of my lack of knowledge about Chinese mountain problems. Thus, the draft recommended two or three reconnaissance expeditions from which a more detailed and long-term programme could then be formulated. This would involve brief visits by small groups to the central Tian Shan in Xinjiang Province, to the northern Hengduan Shan (River Gorge Country) in western Sichuan Province, and the southern Hengduan Shan in northwestern Yunnan.

Huang Pingwei forwarded the draft to his Beijing superiors and, with minor modifications, it became the basis for a formal request to UNU.

Following concurrent exchanges with Tokyo, Bern, and Paris, it was agreed that we should attempt a reconnaissance to the Tian Shan in 1981 and to both the Sichuan and Yunnan Hengduan Shan the following year. The main themes were to be:

- mountain hazards, their mapping and possible mitigation;
- determination of the timing and impacts of deforestation; and
- interplay of minority peoples, population growth, land use, and related downstream effects.

I thought this outline was rather bold, bearing in mind that not only was more than 95% of Chinese territory still rigidly closed to foreigners, but such a programme would require close contact with minority peoples, including Tibetans and Uigurs (Moslem), Naxi, and Yi. Nonetheless, this approach seemed to be following the lines of encouragement that I had received directly from Deng Xiaoping.

© The Author(s), under exclusive license to Springer Nature Switzerland AG 2022
J. D. Ives, *Sustainable Mountain Development*, https://doi.org/10.1007/978-3-030-96029-2_9

Despite this, I anxiously awaited responses from Tokyo and Beijing. It transpired that Walther Manshard telephoned in a remarkably short time to inform me that agreement had been reached and the necessary UNU funds were available to match those committed by CAS.

In the meantime, Roger Barry, Mischa Plam, and I had decided to go ahead with the founding of the International Mountain Society and its proposed quarterly journal. Walther had confirmed that UNU would provide US$18,000 per annum, while UNESCO and UNEP had agreed to cover the cost of four specific issues.[65] Following this, we obtained support from Alpine country colleagues to become members of both the editorial board of the journal to be named *Mountain Research and Development* and officers of the new society.[66] I was able to extend this to include members from China, Japan, India, Thailand, Argentina, and Costa Rica. The most vital factor for the long-term success of the journal was the recruitment of Pauline Ives as its assistant (and more than full-time) editor. With a quarterly journal, I realized we had acquired an efficient system of international communication as well as one that brought in a flow of new research on mountain problems worldwide. The frontispiece for the first issue was a photograph of the Potala that I had taken during the 1980 visit to Lhasa. I prepared a foreword for the first issue which was signed by UNU Rector Soedjatmoko.

The 1981 Tian Shan reconnaissance was planned for the following June. Roger Barry and Gordon Young, a Canadian glaciologist,

agreed to accompany me and we arrived in Beijing in late May.

There was a little more time to make arrangements for the 1982 Sichuan–Yunnan reconnaissance. Bruno Messerli, Corneille Jest, Lee MacDonald (Walther Manshard's senior assistant), and I made up the small visiting team. In preparation, I read all I could about the three large regions we were to reconnoitre. This included the pre-World War II writings of Joseph Rock, whose work became a major element in the eventual research programme that we formulated.[67] Concurrently, Bruno visited the widow of Professor Eduard Imhof, who had led an expedition to the Gongga Shan in 1929–30. Frau Imhof kindly donated the two remaining copies of her husband's book.[68] This gave us invaluable access to photographs and Imhof's watercolours of Sichuan's mountains, forests, and terraced agriculture of a half-century earlier for comparison with our own planned observations.

Tian Shan Reconnaissance, 1981: Mountain Hazards and Environmental Impacts

The plan to travel from Beijing to Ürumqi and on to I'Ning and China's northwest frontier with the Soviet Union was astounding for us at that time. The northwest frontier was rumoured to be the most heavily defended international border in the world. On disembarkation at Ürumqi, we were met by Dr Qui Jiaqi and two of his colleagues from the Xinjiang branch of Academia Sinica. Jiaqi was destined to spend the following

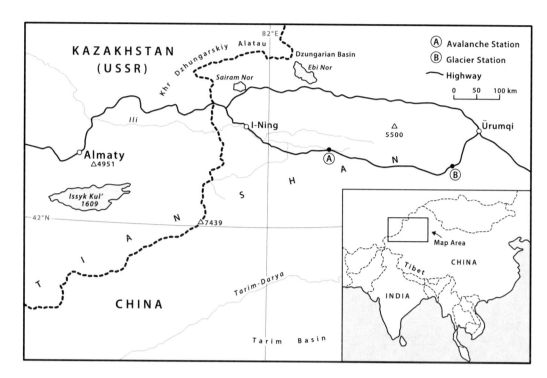

Map 6. The reconnaissance region, Ürumqi to I'Ning, along the Tian Shan, NW China. (with inset of China)

academic year with me in Boulder as a UNU post-doctoral fellow to gain experience with avalanche research and rescue operations. He introduced us to a young Chinese lady, Madam Xiong Jianhong, who was to be our interpreter. Certainly we were dependent on Jianhong as spoken English was rare in this part of China.

The flight from Ürumqi to I'Ning the following morning crossed a wide section of the Tian Shan (The Heavenly Mountains). Because our aircraft was a turbo-prop without pressurization, we flew at a low altitude. As we crossed the mountains, "low" was relative at 5000 m (18,000 ft) and we could see chinks of the rugged landscape through gaps in the floor of the aircraft. The lack of oxygen at that altitude soon became noticeable and all passengers were relieved when we landed at I'Ning.

On the day following our arrival, we were driven on gravel roads northwestward to a high pass in the Dzhungarskiy Alatau range, an offshoot of the main Tian Shan. We viewed the timberline and the dense stands of spruce (*Picea schrenkiana*) interspersed with luxurious green meadows. The landscape was the product of centuries of Uigur and Kazak summer transfer of grazing herds to high altitude (transhumance). Here was one extensive region where deforestation was not a product of the twentieth century, but the result of several hundred years of local pastoralism. We lunched looking out over Sairam Nor, a lake that reflected the western section of the Dzhungarskiy range far into Soviet territory, and we realized that our nearest large town was Alma Ata (since renamed Almaty) in Kazakhstan. A group of extremely curious Uigur children surrounded us. I noticed one

Fig. 58.   Uigur children, initially very shy at meeting their first strange foreign visitors, are soon overcome with curiosity (1981).

small girl with blue eyes. With a modest smile Madam Xiong Jianhong explained that surely we were aware that Ghengis Khan and the Golden Horde had not only passed through the Dzungarian Gate but had reached the gates of Vienna.

For the excursion to the avalanche station, three jeeps and two additional interpreters were assembled. As the supposed leader of the visitors, I retained Jianhong, the most accomplished interpreter and was assigned to the lead jeep, leaving my colleagues Roger and Gordon to follow. Jiaqi rode with me so that, between him and Jianhong, I received an informative account of the landscape through which we were passing.

The total distance from I'Ning to the avalanche station was almost 180 kilometres. With the many photo stops and a visit to a Uigur village, we had a long and interesting day. In the village, we were taken into a traditional household where the couple and two small children greeted us kindly, but with extreme shyness. We assumed this family had been carefully pre-selected and prepped although our welcome appeared completely genuine. I suppose it was inevitable that, considering the remarkable invitation to visit such politically long-isolated territory and people, we were constantly wondering if there were any ulterior motive. We could hardly guess their thoughts.

We proceeded up the valley of the Ili River and into its main tributary, the Gonmeis. The scenery became progressively more mountainous. The avalanche station had a rather precarious location: its out-buildings were

Fig. 59.   A brief pause in a Uigur village entails a formal greeting from its elegant headman.

on the margins of an avalanche path and had been damaged in the recent past. There was a permanent staff of seven, all Han Chinese. Their main duties were to protect the road from closure by avalanches and to maintain a first-order meteorological station. Some of the avalanche defence structures surprised us, as they consisted of hand-constructed terraces on the steep slopes above the road. When we asked how they attempted snow slope stabilization, we were amazed to learn that the large number of labourers who had worked on construction of the terraces were also employed to trample down the snow and so compact it. The Chinese approach to avalanche protection was rather primitive, presumably reflecting the long period that China had been closed to the outside. I could well understand that CAS had readily approved Jiaqi's application to spend a year in Colorado.

The following morning, we continued eastward up a now steeply climbing road for a distance of between 40 and 50 kilometres. We emerged from the mountain spruce forests on to wide open windswept tundra with views into the far distance. The main object of our visit was several kilometres of immense snow baffles that served to protect the road from drifting snow, a huge jump in technology from the avalanche terraces below. The design was based upon wind-tunnel experiments at the Institute of Glaciology and Cryopedology in Lanzhou that would be the final stop of our

Fig. 60. Well above the timberline in the central Tian Shan we inspect the famous baffles erected to combat heavy drifting snow across the open tundra.

Fig. 61. Immense snow baffles designed in the Institute of Glaciology and Cryopedology in Lanzhou. Professor Shi Yafeng, in white hat, Father of Chinese glaciology and lead designer, made a special journey to greet us. Gordon Young stands on his right. Madam Xiong Jianhong wears the conspicuous light blue hat. Qiu Jiaqi, centre, with dark blue jacket and cap (1981).

Fig. 62. While none of the Xi-be archers split the arrow that had commanded the legendary bull's eye, the Sheriff of Nottingham would have had no complaints on this occasion (1981).

tour before returning to Beijing, and Professor Shi Yafeng, innovator of the baffles, had made a long journey from the opposite direction to greet us.

After inspection of the avalanche station, which we considered somewhat basic, and the high mountain road defences, we returned to I'Ning. The following day, we were taken to a small village close to the Soviet frontier. It had been the home of an unusual minority group for several centuries, one of only two such, originally fortified villages set up to check Czarist incursions. Apparently the Xi-bes nation had originated in the extreme northeastern marches of China and had

earned a reputation as fierce fighters, particularly as archers in hunting Siberian tigers. By imperial decree, one half of the population had been transferred to China's northwest frontier, where they had survived with their language, customs, and dress intact. During the previous two decades archery had become a competitive Chinese national sport; the two villages had carried off many prizes, although by then their equipment was decidedly high-tech.

We were greeted warmly and told to expect a display of some of the finest archery in the world, then lunch, and finally traditional dancing. Jiaqi and Jianhong introduced each of us in turn (no doubt, as "VIPs of great importance"). Nevertheless, when they were told that I had spent my undergraduate years at the University of Nottingham (UK), I was startled by a round of applause. I was immediately offered a bow and asked to take part in the archery competition. I was amazed to learn, even in this very isolated village, probably as far from salt water as any in the world, that the legendary fame of Robin Hood was part of the local lore. Of course, I declined the contest, explaining that my intent was to shoot only with my camera.

Immediately after a ceremonial tea, we were escorted to the archery butts. To me, the targets appeared similar to those used in the old Robin Hood movies, although the bows were rather more advanced. The teams were mixed, and both the women and men gave an excellent account of themselves. Although no one managed to repeat the fabled splitting of an arrow, there were many bull's eyes. Lunch was set up in the small village square and followed by a colourful display of ethnic dancing.

On the following morning, we should have flown back to Ürumqi. However, the flight was cancelled due to poor weather. As there would be no plane available for another two days, a road trip was organized. Roger, Gordon, and I were pleased with this as it would enable us to travel along the northern slope of the Tian Shan and across the fringes of the Dzungarian Desert. For me, it was also a chance to see the historic Dzungarian Gate, through which, as local lore relates, Genghiz Khan and the Golden Horde had burst out in the thirteenth century to conquer Central Asia and threaten Europe. Despite the sense of adventure and exploration of a remote region, the road journey became a bone-jarring, two-day 700-kilometre drive in jeeps fiendishly lacking any notion of spring suspension.

When we reached Ürumqi two days later, thoroughly shaken but not immobile, professor Shi Yafeng and several colleagues met us. They had journeyed down from the glaciological station and we all assembled for a welcoming banquet. When I realized that Jianhong was missing, I asked Shi Yafeng why she was excluded. "But she is the only woman!" came the response. I gently objected and she was invited to join us.

I had been warned that, as a guest at a Chinese banquet, one should be careful not to stand while making a toast – and many toasts were to be anticipated. A standing toast carries the imperative of emptying the glass with one gulp, and *moutai* is not to be treated lightly.

The toasts began early in the meal of

Fig. 63.   Nottingham and Robin Hood notwithstanding, our Xi-be lady archers put on a convincing display equalling that of their menfolk.

many courses. We were first welcomed as their only foreign visitors to that date. Of necessity, that was a standing toast. Then I received a personal toast clearly intended to be funny, reflecting the adage that "East-is-East and West-is-West,"etc. It appeared that Chinese children were brought up to believe that Westerners all looked alike (a neat reversal of my boyhood prejudice that all Chinese looked alike). Then we were told that it was with great relief that they learned I had red hair, so they would be able to recognize me on arrival. Much laughter at this point. I had to respond, also standing. I made the obvious retort and the Chinese look-alike was even more relevant because everyone was still (in 1981) obliged to wear "Chairman Mao suits." Then I paused and, looking at Jianhong, commented that, even so, it was possible to distinguish between men and women. From the applause I had obviously made a good point and, at the same time, shown appreciation for Jianhong's presence.

The glacier station, located high in the central Tian Shan with relatively easy access from Ürumqi, was in 1959 the first glacier research station of its kind to be established in China. Despite short interruptions during the Cultural Revolution, under the direction of professor Shi Yafeng it had a long record of glaciological and climatological observations, from what has become known worldwide among glaciologists as Glacier Number One. It was of fundamental importance for UNESCO's efforts to establish an international glaciological network because it was one of a very few glaciers representative of arid Central Asia.

After a steep drive of about 120 km on a narrow, twisting road to the glacier station at 3588 m, we lunched and set off for the glacier. We walked through upland mountain meadows and within little more than an hour we were ascending the glacier's lower tongue. We inspected a long tunnel that had been cut into the glacier and extensively instrumented for recording temperature at various depths and for measuring ice movement. Well aware of the purpose of our visit, Shi Yafeng promised full access to the research station and its surrounding glaciers if we selected the Tian Shan area for our research venture.

The following morning, together with Jiaqi, we descended the steep road back to Ürumqi. In those days, Chinese drivers frequently attempted to save fuel by turning off the engine on steep downslopes. After fishtailing round a couple of hair-raising sharp bends, we agreed that strong action was required and suggested to the driver that, apart from the danger, the cost of losing brake linings and rubber from the tires would far outweigh any savings in fuel. This insistence caused much laughter, but we achieved our aim. Thereafter, the drive was much more relaxing.

We had expected a long excursion the next day, through the mountains and down into the Turfan Depression. However, we had to postpone this as a matter of etiquette when we were informed that Roger and I were both scheduled to give public lectures for students and staff of the Ürumqi August First Agricultural College and the Geographical Institute (Ürumqi, CAS). Roger gave an overview of mountain climatology while I had the chance to describe the work of UNU and outline the beginnings of our re-interpretation of the popular notion that deforestation by subsistent mountain farmers in the Himalaya was producing serious soil erosion, landsliding, and downstream flooding and siltation. We had a large audience and translations, sentence-by-sentence, were by Jianjong. This must have been an ordeal for her as the talks, along with very many questions, lasted more than five hours. Of course, none of our audience had previously heard non-Chinese lecturers and we provoked many penetrating questions. The occasion was a good opportunity to familiarize a large student audience

Fig. 64.   We pause for a brief respite and glaciological discussion on our hike to Glacier Number One. Shi Yafeng is pointing, Gordon Young quietly sits on the rock, while Roger Barry is practicing Chinese and wearing his locally acquired cap and white gloves. Glacier Number One is in the right background.

with environmental and climatological principles from which they had previously been isolated.

Landslides on the road through the Tian Shan into the Turfan Depression prevented our planned visit to this "hottest place on earth." We flew to Lanzhou on 21 June where we were met once again by the convivial Shi Yafeng. We spent time in the Institute of Glaciology and Cryopedology, were taken on a tour of a nearby section of the Loess Plateau and visited the city. The high level of air pollution in the city was beyond belief.

From Lanzhou we returned to Beijing for final discussions about our Tian Shan experience and details of the next year's visit to the Hengduan Mountains in the southwest. Problems of avalanches, landslides, and increasing impacts of intensified agriculture in the Tian Shan area presented a wide range of research opportunities, but we found it difficult to identify a central core project, due to the great distances and the extreme isolation of Ürumqi itself. We thought the Hengduan Mountain reconnaissance would prove more practical, although any such decision would have to wait until we had been there.

This first long journey through some of the more remote parts of China from which foreign visitors had hitherto been excluded reassured us that Deng Xiaoping's invitation of the previous year was to be taken seriously. Perhaps we would have access, both to ethnic minority peoples and Chinese students and scholars hitherto secluded from the outside world. The agreement for Qui Jiaqi to spend a year with me in Colorado would at least ensure continuing contact with Ürumqi.

The Hengduan Shan, 1982: Western Sichuan

The second phase of the Chinese mountain reconnaissance comprised visits to two quite distinct regions: western Sichuan and northwestern Yunnan. Both were areas of hitherto restricted high mountains, home to many different ethnic minority peoples. They were areas where, with the exception of small U.S. air bases in the Yunnan mountains during World War II, no Westerners had visited since the 1940s, and very few before.

Corneille Jest, Bruno Messerli, Lee Mac-Donald, and I arrived in Beijing in mid- September from Paris, Bern, Tokyo, and Boulder. There were introductory meetings in several institutes of CAS, and departure for Chengdu on the 16 September. We travelled under the expert guidance of Professor Li Wenhua, deputy director of the Commission for the Integrated Survey of Natural Resources (CAS).

From Chengdu we began our long road journey, heading southwest across the rich agricultural lowland towards the eastern mountain front. We lunched in Ya'an, gateway to the mountains, then followed the road to a high pass at about 3,000 metres – the Erlang Pass. On the far side, on a precarious twisting and narrow road, we sighted our first high snow mountains in a clearing western sky. Several landslides had crossed the road and had been cleared by bulldozers a short time previously, leaving a surface that degenerated to soft mud in places. Eventually our convoy was halted by a stalled truck. Not only were the rear wheels axle deep, but the driver was struggling under the hood, presumably trying to start the engine. Fortunately we had a very alert driver. He leapt out of our jeep and rushed up to the stalled truck, jumping up beside the driver. Obviously he had anticipated what was happening and managed to throw a heavy cloak over the engine block just as the distraught driver was about to apply a lighted match to the carburetor – no doubt in his mind it was a last ditch attempt to restart his engine. This was one of several incidents when not only did our progress hang in the balance but Chinese drivers astounded us by their responses to emergencies.

A serious accident, however, had been prevented and after much effort, the engine was cajoled back into life, and the truck edged away so we could proceed. We continued our tentative descent into the valley of the Dadu River and to the famous suspension bridge where, at the beginning of the Long March, Mao's advance guard narrowly averted disaster by barely holding it against the approaching forces of Chiang Kai-shek. This had

Fig. 65 [top].   We drove by road across the rich agricultural land of western Sichuan and entered Ya'an, gateway to the northern Hengduan Mountains. We had difficulty driving through the town as the roads were strewn with drying rice and wheat and the walls of the houses were covered with suspended garlands of maize (1982).

Fig. 66. [bottom]   As we descended the western side of Erlang Pass we encountered landslides, but more dangerous was the approach to truck driving displayed by a number of Chinese drivers. The truck shown here was saved from having its carburettor deliberately set on fire by our own quick-witted driver. It is still in difficulty although our small convoy was able to squeeze past (1982).

allowed the Communists to escape from being bottled up for an approaching winter in the mountains of the southwest, a vital turning point in the history of modern China.

The Dadu is a large fast-flowing river and its only bridge for several hundred kilometres had been of crucial importance. A posed group photograph was called for (see Fig. 67).

After nightfall we reached Luding, altitude about 1500 m, and took off again after an early breakfast. We followed the Dadu River northward for about 25 km, then turned sharply west into the broad valley of a large tributary leading to Kangding. The valley displayed distinctive architecture – tight groups of stone houses surmounted by impressive

towers, both for grain storage and defence. We were in the mountainous battleground of formerly warring ethnic groups. Present-day inhabitants were Tibetan, their subsistence farming a mixture of yak and sheep grazing and terraced cereal cultivation.

Only the lower terraces were being used for barley; most of the higher terraces had been abandoned for some time. Bruno and I were both quick to remark that there appeared to have been little change since Eduard Imhoff had made his watercolour and photographic records more than 50 years ago. We replicated several of Imhoff's scenes with our own cameras. Here again we questioned the assumption that the environmental damage

Fig. 67.   An improvised pose on the famous bridge across the Dadu River (1982). The bridge was seized by Mao Zedong's vanguard hours before Chiang Kai-shek's forces were able to destroy it and so prevent the Long March. Bruno Messerli, Corneille Jest, and Lee Macdonald stand on the bridge with several of our Chinese colleagues. Professor Li Wenhua is on the extreme right holding onto the handrail.

Fig. 68.   Upstream of Kangding we enter a broad valley with agricultural terraces little changed from 1930 when photographed by Dr Edvard Imhoff with the Swiss expedition to the Minya Konka (Gongga Shan). Note the defensive architecture and the barley cultivation confined to the lowest terraces (1982).

– deforestation, landslides, and soil erosion – had occurred during the last 50 years, a thesis that had been applied especially to the Nepal Himalaya. And the opportunity to replicate photographs taken more than a half-century ago as a means of checking landscape change made me realize that we should promote repeat photography as a major tool for research – thus began my systematic search for relevant old photographs.

The road climbed slowly westward until we reached Kangding at an altitude of about 2600 m. We then turned almost due south and drove for another ten kilometres before Li Wenhua announced that we would be transferring to horseback. We were soon surrounded by a large group of Tibetan horsemen with numerous spare mounts. There were hearty greetings from our new guides who reminded me of scenes from old issues of *National Geographic* magazine – fierce faces, outlandish garb with broad-brimmed hats and lassos, and with unforgettable smiles. They were nomads employed to provide transport and general support for the CAS scientific base camp. We were later to meet many more, including their womenfolk and children with whom we would be camping that night.

Next came the business of mounting us; our exotic horsemen laughed uproariously as they tried to hoist Bruno, Lee, and me into our saddles (Corneille, of course, was quite adept). The problem was that the stirrups were far too short, and the saddles, made with wooden frames covered with blankets, left us uncomfortable from the beginning. My Hasselblad was allocated its own horse. It was a

Fig. 69.   Shortly after passing through Kangding we came to the end of the road where we were met by Tibetan nomads and more than two dozen horses. Our new leader and chief guide agreed to pose for a photograph.

thrilling beginning with ten in our party and about a dozen Tibetans and many spare horses. We set off amidst shouts and hurrahs from our seemingly ruffian guides with about ten kilometres to go. It would have been a sight to photograph, but the Hasselblad was already tightly lashed down.

As we progressed, with knees up high and the boards of the saddle beginning to bite, it became a struggle to keep going. We eventually arrived. A crowd of women and children streamed towards us with cries of welcome. I felt certain that bones were out of place and I needed help to dismount. I could hardly stand, let alone walk. The camp doctor was called. A young Chinese medical officer greeted me with great concern. A quick examination was followed by a question: Would I prefer acupuncture or a deep massage? I chose massage. I was asked to lie on my stomach on a rolled-out mattress. He hoisted my shirt over my head and pulled down my pants to my knees. Then, kneeling astride me, he set to work. We were quickly surrounded by a host of laughing women and children. I was mortified. Presumably they had never seen such white buttocks before! But the luxury of the massage quickly banished any sense of embarrassment. After the pummelling, a "hot" poultice was strapped around my waist and hips. The doctor explained that it was a superior mixture of mustard and ground up tiger bone, but it must be removed before breakfast next morning or my skin would be burned!

The altitude of the camp was about 3800 m. We slept in Chinese sleeping bags inside roomy black felt tents. Despite the altitude, I needed my tent door open, for no doubt the ground-up tiger bone was doing its job. I quickly fell into a comfortable sleep and awoke in time for breakfast, feeling decidedly cosy. I cautiously crawled out of the sleeping bag to find, to my astonishment, that I felt more flexible than I had been for years. The dressing was removed, and I sat down to breakfast comfortably cross-legged. I was soon ready for the horse ride up the neighbouring valley to the Zimeiling Pass where we would view the Gongga Shan (or the Minya Konka, as known to Edvard Imhoff).

We meandered up a rather steep valley under a disappointing sky. Nevertheless, we reached a broad open pass just before noon and were able to look along the western flank of the high mountains although there was much low cloud. Then there was a sudden clearing, and we could see the majesty of the great mountain through a thin mist. I managed to take a photograph, but when I looked again, the view we had travelled so far to see was lost.

We descended slowly. Bruno, Lee, and I decided that walking was more comfortable than horseback. After lunch we prepared for departure. Then there arose an impasse. Our Tibetan horsemen were demanding double the fee for the return journey, causing an outburst of temper from our Chinese friends. We were in no position to argue, especially when there was a strong possibility that Tibetans and horses might leave. At this point I intervened: here was a chance to pass on some financial support to our minority friends. I told Wenhua that the UNU part of the budget would cover the extra charge. A quick translation followed and everyone was smiling once again. We set off at a fair pace to the waiting jeeps and were soon on our drive back to Luding.

The final leg of our journey back to Chengdu was uneventful except for one incident. We were caught by a landslide that crossed the road and edged our jeeps towards the Dadu River. This required some remarkably adept driving. While several of our companions, with local helpers, propped up the downslope side of the jeeps to prevent them rolling into the river, our drivers made cautious headway until they reached stable ground. I walked behind.

Reconnaissance to Lijiang and the Jade Dragon Snow Mountains

On 22 September, we caught CAAC 4421 from Chengdu at 6:55 a.m. for an uneventful flight to Kunming. After one night at the Green Lake Hotel, we made a dawn start for Lijiang, estimated to be a two-day drive. Our transport was a large van, and our group consisted of the four Western visitors, Li Wenhua, and three academy scientists, two of whom were biologists based at the Kunming Institute of Biology. There was an additional, strangely silent Chinese gentleman, Mr Hu.

Heading westward, we quickly left behind the sprawling city of Kunming and its traditional picturesque low houses and narrow streets. Our drive for the day was more than 300 kilometres on rather poor roads to Xiangyan. This was an ugly industrial town but only a short distance from Dali, a beautiful mediaeval town that has since become a major tourist destination. Dali is the walled capital of the Bai nation, and in the 1990s it underwent a massive rejuvenation.

As we proceeded westward, we slowly

Fig. 70 [top].   A small remnant of the near-natural forest cover of the Yunnan Plateau northwest of Kunming. The sub-tropical mixed forest of many different tree and shrub species has been extensively cleared (1982).

Fig. 71 [bottom].   In 1982 re-afforestation of the denuded plateau was desultory at best, in this instance restricted to planting eucalyptus along the main road. Wherever soil has been preserved on the valley floors, intensive agriculture is practiced.

gained altitude and entered a progressively more stark, extensively deforested area of the Yunnan Plateau. The valleys still supported rich agriculture that contrasted with the man-made desert of the hills. The bare ground of the hills became more pronounced as our altitude increased. We discussed the prevailing harshness of the landscape and were consequently introduced to what we came to refer to as China's version of the theory of Himalayan environmental degradation – post-1950 devastation of the forest cover. In this instance, however, the explanation rested on the massive exploitation of the forests by Mao Zedong's "Great Leap Forward," the Cultural Revolution, and the "Gang of Four."

Only in the last few years, with the rise to power of Deng Xiaoping, had determined efforts been made to reverse the seemingly inexorable process of deforestation. For us during this first day, driving westward in Yunnan, the only "reforestation" apparent were the lines of eucalyptus along the highway, occasionally double lines. For considerable distances, not only had all forest cover been removed, but erosion had taken most of the topsoil, exposing the bedrock.

The next day, there was a long stop just north of Dali. We had bypassed the mediaeval walls and great fortified gate of the old town and transferred to three jeeps that were awaiting us. We were then driven along tortuous forest roads up the flank of a southerly offshoot of the Hengduan Mountains that reached to about 4500 m. The drive was through a healthy plantation of *Pinus yunnanensis*, a conspicuous mono-species reforestation seeded from the air. The trees were almost uniform in height at four to six metres, smaller on the higher slopes. Far below was a great expanse of farmland leading gently to the margin of Erhai Lake. The reforestation had been very effective, although at lower levels success was limited due to overgrazing closer to the farms.

After returning to the van, we proceeded north, parallel to Erhai Lake, to a large village where we had been invited to stay for lunch

Fig. 72. Much of the Yunnan Plateau has been clear-cut and, over the centuries erosion has removed much of the soil cover to expose bedrock. We referred to this type of landscape euphemistically as Yunnan's 'Painted Desert'. The deforestation is not recent, dating from the birth of Communist China as was widely claimed at the time. Rather, it is the product of centuries of tree-felling and over-grazing (1982).

and join a local Bai festival. This was our first close contact with one of the main ethnic groups of the region. Costumes were very colourful, except that the younger males wore Chairman Mao dark blue or green suits and peaked caps, in contrast to the women, children, and older men. We intermingled freely with the villagers and took many photographs of the celebrations.

Progress during the afternoon was very slow as the scenery became more rugged. As we passed through small villages, people turned out en masse to greet us. The road led us through a prominent mountain ridge, following the side of a deep gorge. We would remember this section on our return.

Farther north, we passed through broad depressions set amongst mountains, and more numerous villages where we noted that the local dress was much less vivid than that of the Bai people. The subdued dark blues and whites indicated that we had entered the territory of the Naxi, an ancient people who claimed that their written language (pictographic) predated that of the Han. At one point, the Naxi farmers had spread out along the road their recently hand-cut rice. By driving over it, we helped separate grain from the husks. Our unintentional assistance earned us waves of thanks.

Our first view of the Yulong Xueshan (Jade Dragon Snow Mountains, highest summit 5590 m) appeared at last, indicating that we were approaching Lijiang, our destination. We were housed in the only hotel considered acceptable for Western visitors. Wenhua apologised for the absence of running water, minimal restaurant arrangements, and

outdoor toilets that were best used unseen after dark. Jugs of water and hand bowls in our rooms were quite adequate. Our anticipation of being in the land of the Jade Dragon, home of the strange Dr Joseph Rock for most of the period between 1922 and 1950, more than compensated for any real or perceived inconvenience.

Early breakfasts had become a routine part of the entire Chinese mountain

Fig. 73.   Our first introduction to the colourful costumes of the Bai people. This photograph demonstrates my preference for female subjects. But in this instance, the young women were initially facing me. As they saw the camera they turned away, not realizing they had exposed the most beautiful view of their costumes.

reconnaissance undertaking. We left Lijiang and drove north along the eastern flank of the Yulong Xueshan. Outside the limits of the Old Town (Dayan), we passed through indifferent rows of houses as the town had increased in population and spread beyond the walls already before World War II. The Lijiang plain stretched northwards for about 30 kilometres. At its southern, Lijiang end, it was about ten kilometres wide with steep western slopes that reached upward to glaciers and permanent snow fields although because of the cloud cover all we could see were the lower forested slopes. On the east side a low ridge, running slightly west of north, had its crest in full view. Its slopes were forested to the ridge crest, and steep gashes from apparently recent erosion cut through periodically to reach the farmland of the plain.

The plain supported a rich agriculture,

Fig. 74.   Our 1982 reconnaissance party in the Love-Suicide Meadow north of Lijiang. The mysterious Mr Hu (Who?) is on the extreme left. Lee MacDonald stands next to him with Bruno, Corneille and Wenhua close to the centre of the group.

and the road, more or less straight and due north, avoided the villages. The plain narrowed progressively northward, and eventually our route climbed a steep slope and we entered a succession of small tectonic basins then descended steeply into the "White River" valley. We crossed the turbulent river on a dilapidated stone bridge and turned into the courtyard of the local headquarters of the Forestry Service. This was to be the farthest point for our van, which we left in the courtyard and proceeded on foot into the forest.

A well-used trail led us steeply upward through a dense forest of large trees – evergreen oak, conifers, and deciduous species. After about four kilometres, we broke through the forest onto a large meadow at about 3000 metres. Three years later, on my return, I learned that this was the mysterious "Love-Suicide Meadow" of Naxi folklore. I was surprised that Wenhua had not informed us; perhaps he had been unaware of its significance.[69] We were told that Joseph Rock had camped in the meadow on several occasions. After lunch we retraced our outward journey back to Lijiang.

So far, we had learned that the local ethnic groups were a potentially vital source of information. The older people informed us that the pattern of deforestation extended back in time, at least to their childhood and probably much longer, in contrast to the "modern" slant of the day. Their agricultural systems, costumes, and general way of life appeared to have stood up rather well to the abuses of the Cultural Revolution, although they had been collectivized and their farming and forestry systems constantly modified by Communist

Fig. 75. Yuhu village and the farm in 1982, the upper story of which served as Rock's headquarters for most of his prolonged stay in northwest Yunnan. The man on the left side of the small group was taught to read in English by Rock when he was a small boy.

Party dictate. They would constitute a vital source of information and understanding if we selected this region for our UNU project.

Since our arrival in Lijiang, our hitherto almost silent friend Mr Hu had come to life with numerous attempts to amuse us. His English was quite good, although he avoided all questions concerning his affiliation and functions, limiting himself to such statements as "I am here to help." Corneille suggested to Bruno that they speak to each other in dialect if they wished to make any private comments. Bruno's Bernese matched Corneille's Alsace dialect sufficiently for mutual understanding but was unintelligible to anyone else. We began teasing Mr Hu by asking, "Mr Who, who are you?" We thought he was an agent of the secret police and that the Bernese/Alsace dialect would have stumped him. His actual status was confirmed later (see below).

The second and third days were spent amongst the villages on the western edge of the Lijiang plain, where we were introduced to

Fig. 76. The Naxi lady in traditional costume is holding a framed group of photographs, including colour, of her parents generation taken and presented to them by Rock. She explained that she had protected these heirlooms from destruction during the Cultural Revolution by hiding them in the walls of the farmhouse.

Naxi rural culture that appeared little changed over centuries despite the rigours of collectivization. A mixed farming culture predominated with Lijiang providing a local market. The impacts of heavy logging were evident everywhere and, although we were told it had been made illegal, it was obvious that logging persisted.

The final village on our "checklist" was Yuhu, the location of Joseph Rock's former headquarters. We were shown the actual farm where he had based his Yunnan operations for several decades. It was the upper floor of one of the larger farms. The present inhabitants had lived there for several generations. They showed us family photographs Rock had taken of them that had survived the Cultural Revolution because they had been concealed inside the walls of the farm buildings.

Our final day included a visit to the Old Town and the city park. The name Lijiang, we were told, means "sweet water"; centuries ago, the Lijiang River had been diverted into a series of canals around which the town had grown. The water had been clean enough for drinking; in fact, it was reputed that the inhabitants used to fish for trout from their garden gates. While somewhat run down at the time of our visit, it was, nevertheless, highly picturesque although the "sweet water" was no longer potable.

I was especially attracted to one of the city's main gate-houses, which I photographed. This was fortunate because, despite several later visits between 1985 and 1995, I never saw the gate again. I presumed it was demolished during a clean-up in preparation for mass tourism after 1985. We had been privileged to see Lijiang as it was when Rock had lived nearby.

In contrast to our northern journey, on the return we saw many heavy trucks, heading towards Lijiang. We ascended the steep narrow road cut into the gorge side and exposing a precipitous drop-off down to a mountain torrent several hundred metres below. As each truck passed, our careful driver reduced speed and pulled over to the edge. Presently we approached a blind corner. A few moments later, a truck rounded the corner, followed instantly by another that began to overtake it. The second truck was pulling a trailer loaded with long metal reinforcing rods and it totally blocked the road. Instead of braking, the second driver accelerated, assuming that he could overtake in time to pass us safely. We tensed as we realized that a potentially serious accident could hardly be avoided.

Fig. 77.   The old gate house of Lijiang (Dayan) as seen in 1982 but never again, due to its most regrettable demolition.

Our driver pulled over to the very edge and stopped. We waited for the impact. The first truck, sensing the danger, braked while the second continued to accelerate and tried to squeeze between us. He hit the first truck with considerable force. The impact caused release of the reinforcing rods as high-powered projectiles, and five of them shot through our windscreen, coming to rest amongst us and shattering glass throughout the van. It was remarkable that no one was hurt or that the impact had not sent us over the edge.

No one reacted. We all sat silent, breathing deeply – that is, except for Mr Hu. He leapt out of the now badly damaged van, immediately accosted the offending driver, and arrested him. He took particulars from the first driver. The two trucks were jammed together, and it appeared that machinery would be needed to separate them. Mr Hu took out a small radio and sent a message. Within 30 minutes, four police cars appeared. One stayed with the wreckage; the others transferred us to their headquarters some 15 km farther south. Within a short time, a second van arrived and we were whisked away as if nothing had happened. Corneille took the occasion to shake Mr Hu's hand and thank him: "Now we know who you are, Mr Who. We are very glad to have had you with us."

We followed the same outbound route southward and entered the Bai village where we had stopped on the way north. Here we were directed into a walled courtyard through a high reinforced gateway. The gates were shut and locked behind us. In total contrast to our earlier reception, we were met, very severely, by a group of village elders. Wenhua, Mr Hu, and I were asked to accompany them into a hall where we were formally seated. The tension was palpable. That something was seriously amiss was very clear. A conversation ensued and it was apparent that it was not friendly. Finally, a somewhat disconcerted Mr Hu gave me a summary of the problem. During our stopover on the way north, one of us, despite the hospitable reception, had photographed a villager who was mentally retarded and had been clowning along the street. They assumed that, as we were all Americans, the photographs would be used to mock their poor village by being published, for money, in Western newspapers. The village headman insisted that all our cameras and exposed film be confiscated.

It was immediately apparent that loss of cameras and film was a small part of our problem. I had already decided to recommend the Lijiang area for future detailed research. This present problem somehow had to be solved and I had no idea what to do. So I apologized and asked for a chance to talk with my colleagues. After some discussion, I was allowed to leave but told to return within half an hour.

I outlined the disaster to Bruno, Corneille, and Lee. Lee courageously tried to explain. He had been accosted in the village crowd by a young man whom he thought was merely making a display in order to be photographed. He had no idea that the man was mentally impaired. So that must be given as the explanation. We quietly talked the situation over. It was agreed I should give the straightforward simple explanation and offer one of Lee's exposed rolls of film to the village headman.

First, I explained to a hostile audience that we were not all American – one French, one Swiss, one Canadian, and one American. I felt like a heel saying this. Then I explained that the photographs were not taken with any malicious intent and they would never be offered to the newspapers. Mr Hu translated. After some discussion, he turned to me and said my explanation had not been accepted. I felt highly perturbed. Then I had a brain-wave. I asked if the Bai people knew the Sufi story about the blind men and the elephant – about how each blind man was allowed to touch a different part of the elephant and asked to guess the name of the animal. This was really a desperate try. To my amazement and relief, they acknowledged they knew the Sufi story and were surprised that someone from "America" did. I noticed a few half smiles as Mr Hu was translating the story and realized that a barrier had been broken. I followed up what I saw as a small advantage and likened their four visitors to the blind men standing before the elephant. I explained how privileged we felt in being allowed into their country, the first Westerners since before 1950, and how we were following in the steps of Dr Joseph Rock (with whom they were also well acquainted). I offered as a token of our faith that I would leave with them the exposed film.

To my immense relief, the village headman came around the table and embraced me. Everyone began talking at once and I was asked to tell the others that we were to be their guests for a meal. Mr Hu explained that my relating the Sufi story had proved the turning point. I staggered outside. My appearance was enough to bring smiles of relief to the faces of my friends. But I asked Lee for his exposed roll of film as I thought it imperative that this be used as the article of trust. We had a substantial meal with much local fire water, fortunately in small glasses. Half the village turned out to wave us goodbye, including the poor man who had been the centre of all the commotion. I needlessly joked with Lee that he was not to take any photographs.

Our final day, October 1, was a national holiday and Kunming was prepared for a great celebration. The morning was spent touring the western cliffs with their ancient temples, lookouts, and walkways. In the afternoon, we visited the Yunnan Centre of CAS and met several members of the Institute of Biology where we exchanged views and sought suggestions on future research. The institute staff had undertaken research in both the Dali and Lijiang regions that included extensive mapping of the flora, involving mountainside traverses from the lower subtropical belt to timberline and alpine tundra at the highest levels. This would be important for questioning the assumed massive impacts of deforestation. The institute staff were enthusiastic about our plans and were willing to provide a senior botanist to work with us.

On 2 October, we returned by air to Beijing where we spent three days. We visited several of the CAS institutes, including Geography where we were welcomed by our old friend Professor Huang Pingwei, and the Institute for the Integrated Survey of Natural Resources. Sun Honglie was the director and he had set aside a day and a half for informal discussions. This proved an excellent forum

for discussing the results of our three reconnaissance expeditions.

It was agreed that the Lijiang-Yulong
Xueshan was the most suitable choice for
development of a long-term research programme, provided the authorities would be
supportive of detailed questioning of the local
people about their land-use systems, their approach to forestry, and their knowledge of the
history of forest cover change. We indicated
how such a study would parallel the work already begun in Nepal and explained our questioning of the claims for recent catastrophic
deforestation. After highly productive discussion, we agreed to develop a joint research
proposal that CAS would submit to UNU.

Halfway through a response by Yongzu, who had already spent a year with me in
Colorado, I was surprised to feel a firm hand
on my shoulder. Here was the mysterious
Mr Hu again. Would I please go with him?
Much surprised, I quietly left the room, to be
hustled downstairs to a waiting limousine. A
uniformed driver with sparkling white gloves
opened the rear door for me, while Mr Hu entered through the front passenger door.

Eventually we drew up in front of a large,
distinguished building; at least it didn't appear to be a prison! Mr Hu led me into the
building and along a wide corridor to an ele-
vator. On the first floor, we entered a large
drawing room full of a dozen or so serious-
looking gentlemen seated in big armchairs.
They all rose as Mr Hu presented me to Professor Ye Duzheng, the vice-president of CAS.
Professor Ye, in turn introduced me to Mr Hu
Yongchang, deputy secretary-general, and invited us all to be seated. I noticed, also to my

surprise, that Honglie was present. He must
have left Yongzu's presentation about the
same time I did, but travelled by a different
route. I took the conspicuously empty chair
between the two senior gentlemen and waited as tea was served.

Several cups of tea were consumed in
stately near-silence followed by a clear indication that I was expected to say something.
But on this first occasion, what was expected
of me? Not the story of the elephant and the
blind men! I assumed that an account of my
impressions of our highly privileged travels
was expected – of the previous year's reconnaissance to the Tian Shan and our recent visits to Gongga Shan and northwestern Yunnan.

I paused, reached for the remains of the
final cup of tea, and offered a short account,
concluding with an expression of gratitude
for the privilege. There were several pointed
questions, some requiring the assistance of
Mr Zeng, a high-level interpreter, who later
became a good friend.

There was a pause for more tea. Eventually Professor Ye addressed me with a simple
but, under the circumstances, formidable
question: Who did I consider that I was working for? This was becoming a little beyond my
comprehension. Nevertheless, I explained
that I was a Canadian citizen, born in England (which he obviously would have known),
and although I was teaching at an American
university, I considered that I was working for
the United Nations University where my full
allegiance lay. Professor Ye rose and came to
shake hands and thank me. There followed a
warm farewell from the entire group, Mr Hu
took me down to the waiting limousine and so

back to the hotel. As we alighted, Mr Hu quietly said, to my continuing astonishment, "You have passed the test, now you will be able to travel anywhere in China."

I related my experience to my colleagues to mixed response. Corneille, our worldly anthropologist, couldn't resist pointing out that this was CAS routine. I think Bruno and Lee were more than a little impressed. Later, Corneille approached me with a word of advice: "Jack, I don't doubt that they will invite you to travel across Tibet. You must refuse if they do, because they will use you as a political tool." I decided that the advice was the result of Corneille, a distinguished Tibetologist, being jealous. Four years later, however, he would be proved correct. And so ended our reconnaissance missions into the Chinese mountain hinterlands.

Reflections on the China Connection

The invitation for a small group of Westerners to visit some of the remotest mountain regions of the People's Republic of China in those very first years of openings to the West was remarkable. Furthermore, given the remoteness of the Tian Shan and the Hengduan Shan and the very limited communications development of the time, our CAS colleagues put together an impressive display of organizational expertise: the timing of our meetings with guides and informants along the way had been flawless. And when our van was badly damaged on a mountain pass south of Lijiang, an alternate vehicle was at hand within an hour. We were able to meet with local authorities and mix with minority peoples in an atmosphere of openness and enthusiastic welcome. Finally, my formal interview with the CAS vice-president, Professor Ye, and his senior colleagues in Beijing, clearly demonstrated a convincing level of trust and a sense of commitment to long-term research collaboration with UNU.

Our reconnaissance visits were the direct outcome of Deng Xiaoping's promises relating to the initial opening of Tibet to foreign scholars in 1980. It remained to be seen whether we could use the opportunities both to help advance general knowledge of China's little-known mountains and to apply scholarly findings to benefit local minorities and their environment.

Sun Honglie, on behalf of CAS, as well as Professor Manshard, on behalf of UNU, encouraged me to prepare a draft research programme for evaluation by the two institutions. All three of the regions we visited – the central Tian Shan, western Sichuan, and northwestern Yunnan – had outstanding research potential. However, the Yulong Xueshan-Lijiang region of Yunnan seemed the most attractive by far. It was the most accessible of the three; there was an abundance of documentary material; significant biological research had already been accomplished by the Kunming Institute of Biology; and the photographic legacy of Joseph Rock should be available in the archives of the National Geographic Society. Finally, the initial response of the Naxi people, an essential aspect for any future research, had been most encouraging.

I completed a draft proposal soon after returning home to Colorado early in October

1982. Copies were sent to Bruno, Corneille, and Walther Manshard. Their suggestions were incorporated, and the slightly adjusted version went to Honglie in Beijing. A series of studies were proposed: in geomorphology (soil erosion); the pattern of deforestation (including vegetation transects); and an assessment of the human impacts on the local environment over the last century. These studies would parallel the research programme in the Nepal Himalaya. In particular, they would widen our challenge to the core environmental paradigm of the day that we had begun to describe as the theory of Himalayan environmental degradation. It seemed that the Yulong Xueshan-Lijiang area offered a valuable Chinese comparison, but with intriguing differences relating to the internal policies of Mao's Communist regime.

The photographic record of Joseph Rock dating from the 1920s and 1930s would form an excellent base for replication and determination of any landscape changes over a significant timeframe. Bruno and I had already noted the value of Edvard Imhoff's photographs and water colours of the Gongga Shan region dating from the 1930 Swiss mountaineering expedition. Rock's photographic record was much more extensive.

Any research in the Lijiang area would require detailed knowledge of the Naxi and Yi people, their agricultural systems, standards of living, and relations with the local Han people. We would need to ascertain in some detail the human–environment balance and, if possible, any changes that had occurred during the last 50 years or so.

The next step, therefore, would be a 2–3 month period of fieldwork by a joint UNU-CAS team of researchers and support staff, perhaps as early as April–June 1984. The Western component should include two senior scholars and two doctoral-level graduate students. I invited Barry Bishop, who at that time was chair of the NGS Committee on Exploration and Research, an Everest summiteer in 1963 and an avid photographer. He promised to search the Joseph Rock section of the NGS photography archives and produce high quality prints for replication in the field.

It was expected that CAS would select six scholars, together with drivers, cook, guides, interpreters, and local Naxi liaison staff. This first full-scale venture, while still of an exploratory nature, should produce publishable results. The proposal met with full approval from CAS and UNU. However, the starting date was postponed for a year to provide time for CAS to prepare a substantial base camp and to take care of the many complicated local arrangements that included obtaining formal permission from several levels of government. Barry had to withdraw at the last minute although he shipped by express delivery to Boulder, Colorado, 25 enlargements of the Rock photographs.

Fig. 78.   The perfection of the agricultural landscape of Nepal's Middle Mountains impressed us: seeming stability and careful management. Great flights of khet terraces descend hundreds of metres down to the valley floor. The young rice in the foreground has barely emerged through the irrigation water on the flooded terraces. The water is carefully controlled by low earthen bunds, the surplus being guided down to the next lower terrace by stone-lined cuts in the bund, and so all the way to the river in the valley bottom (1986).

CHAPTER 10

Breakthrough: Environmental Degradation Theory Overturned

The United Nations University project on highland-lowland interactive systems obtained some of its most significant scientific results from the mapping of mountain hazards and investigating the causes and timing of deforestation in Nepal. As indicated earlier, we were fortunate in being able to link UNU, as the primary source of funding and intellectual support, with UNESCO MAB-6 and the IGU Commission on Mountain Geoecology. Together with responsibility for supervising the fieldwork in northern Thailand and the reconnaissance expeditions to China, this proved a heavy workload and the Bernese contributions were indispensible. In addition, we were involved in organizing a series of mountain conferences and associated field excursions in Switzerland and in far distant mountain regions.[70]

Nepal's Middle Mountains: Getting to Grips with the Real Problems

The road access to the Kakani ridge from Kathmandu proved a great asset, and full-scale fieldwork was underway within a few months of completing the 1979 March-April reconnaissance (Chapter 7). Although Bob Kates had been unable to join us, his former doctoral student, Kirsten Johnson, proved an effective alternative. With her Clarke University colleague, Annie Olson, and Sumitra Manandhar, a sub-group was formed that won rapport with villagers in the Kakani field area. This enabled us to better understand the role of the women in poor farming communities where norms of male–female equality and inequality varied amongst the different ethnic groups (Johnson et al. 1982; Manandhar Gurung 2005). The menfolk also were willing to discuss with our entire team their problems and especially their attitudes to the landslide and fuelwood difficulties. It became apparent that many of the ongoing Western assumptions were not supportable. In this we learned much from the concurrent studies of the staff of the Nepal-Australia Forestry Project (Mahat et al. 1986a, 1986b, 1987a, 1987b). Sumitra Manandhar was so committed that she took a course in elementary medicine to equip herself as a village paramedic, with a small clinic and a limited range of drugs. She joined the village women in the arduous task

Fig. 79.   This view looks onto the run-out area of a former landslide. But in this case the farmers deliberately engineered the landslide by diverting a small stream and saturating the slope. By this means, it was much easier for them to carve new terraces for paddy rice (khet), a laborious task at the best of times (1980).

Fig. 80.   Perhaps it is rather the 'ignorant' western engineering advisor who should be credited with causing damage. Here, the Trisuli Road acts as a river bed during heavy monsoon rainstorms. A sharp bend in the road causes the flood water to discharge down the hillside and destroy the poor farmers' subsistence crops and carefully constructed terraces. In the dry season, the contrast between the rain-fed terraces (bari), here bare, and the khet which are being used to grow winter wheat following summer monsoon paddy, is stark (1980).

Fig. 81.   A young Kakani farmer with his special implement for winter-time trimming of the terrace frontal areas (1978).

Fig. 82.   Heavy monsoon rains certainly cause much damage, despite intensive efforts by the farmers to protect terraces from collapse. Burrowing rodents also penetrate the frontal areas. During the winter much effort is expended to clean-cut the terrace fronts (here the marks of the cutting tools are clearly visible). The terraces are fertilized using all available farm refuse as well as forest litter (1978).

of transplanting rice. Subsequently, her contributions to the project led to a doctorate from the University of Hawaii in 1988.

Hans Kienholz and the group of graduate students from Bern University undertook the core of the field mapping work. Nel Caine, my colleague from Boulder, Colorado, assisted by Pradeep Mool, studied the types and frequency of mass movements, linear (river) erosion, and flooding throughout the area. Also of Bern University, Professor Tjerk Peters was our soils expert, and Professor Erwin Frei tackled the bedrock geology. There was no better person than Hans to lead the hazard mapping project, both with his practical and academic abilities and Swiss Army training. He had spent a year with me at the mountain research station in Colorado so that rapport was instant.

The initial field season extended from October 1979 until March 1980. It was followed by a second, longer field season from October 1980 until April 1981 that enabled all the basic mapping and taking of questionnaires to be completed. The map production process was remarkably efficient. Two maps on a scale of 1:10,000, "Land Use" and "Geomorphic Damages" were printed in Bern in 1981. Two more, "Base Map" and "Mountain Hazards and Slope Stability" were printed in early 1983, and a final map, "Categories of Land Tenure," was made available in 1986. The maps were published as inserts to explanatory monographs in *Mountain Research and Development* (Kienholz et al. 1983, 1984), together with a series of related technical papers (Johnson et al. 1982; Caine and Mool 1981, 1982).

The 1:10,000 topographical base map had been essential for the success of our work.[71] The five thematic Kakani maps represented a "first" in Himalayan research. The ability to collect the extensive field data and to produce such maps, along with information on soils, bedrock geology, climate data, and village practices led into a far fuller understanding of the problems facing the Himalayan (Nepal) Middle Mountains than would otherwise have been possible. However, the scale of 1:10,000 was impractical for a county-wide mapping programme, and topographic maps of that scale were limited to the immediate area. Nevertheless, it was an ideal research tool that greatly improved our understanding of the human–natural inter-relations in the evolution of steep mountain slopes in a monsoon climate.

Some of our early findings were reinforced by information from the farmers of the Kakani ridge. They well understood the slope instability problems. The extensive flights of irrigated and rain-fed terraces (*khet* and *bari*, respectively), carefully engineered by generations of farmers, were highly effective soil conservation works. We saw that their management of irrigation water, in general, and heavy monsoon downpours, in particular, showed far more understanding and intelligence than that displayed by many of the so-called experts of bilateral and UN aid agencies. The farmers understood landslide mechanics, within their own vernacular, well enough even to divert mountain streams to cause landslides (debris flows) for ease of terrace construction in the soft runout material.

This did not mean that there was no

environmental deterioration, nor crises following monsoon downpours when entire flights of terraces slumped, taking houses and human beings with them. The farmers of the Kakani ridge, and much of the entire belt of Nepal's Middle Mountains, were living close to the margin. Nevertheless, in the words of Michael Thompson, the farmers were not the problem – they were a major part of the solution(s) (Thompson et al. 1986).

In contrast, we were dismayed by reports of the "experts" of the time. Specialists with the Asian Development Bank, for instance, castigated the "ignorant" farmers for engineering the rain-fed terraces to slope outward from the mountain side. They recommended that they should slope inward and so retain rainwater. This unfortunate recommendation was published in an Asian Development Bank report on Nepal in 1982, directed by Western "experts." Any such approach would accelerate the landsliding process; the traditional outward-sloping terraces were deliberately engineered by the villagers to ensure that the rainwater would run off rather than pond on the terraces and so augment slope saturation, a potent precursor to landslide release.

In 1978, I photographed a recent landslide on the Kakani ridge and marked the camera station for subsequent replication. At the time, I was not sure whether I would be setting up a visual record to demonstrate acceleration or stabilizing of the landslide scar. Several years later – and ultimately 15 years after several replications, these photographs became iconic images. They demonstrated how the local farmers were first able to contain the damage and later ensure that the area

was reterraced and its productivity restored. The local villagers understood that one of their most serious problems was the protection and repair of the main irrigation and drainage canals that criss-crossed the mountain slopes.

On presentation of our maps and papers to the UNDP Resident Representative in Kathmandu, I made a plea for financial support for reinforcement of the main irrigation and drainage canals. I recommended this would require an engineering survey, to be followed by a supply of reinforced concrete. First, I was told that the local farmers would have to provide volunteer labour as such a project would be for their own advantage. My response was that they should receive fair salaries as compensation for their time because subsistence farmers had to work full-time to provide food for their families. I was then asked for a rough initial estimate of costs. I replied – perhaps somewhere between US$150,000 and $200,000. I was discouraged to be told that such a small amount would involve as much bureaucratic paper work as a project of some millions and so was unthinkable. I was so angry that I clenched my fists – and put them in my pockets – as I left the senior bureaucrat to his luxurious office.[72]

With the fieldwork in the Kakani area well in hand and publication of the results proceeding apace, we began to plan for the second of the three parts of the Nepal Mountain Hazards Mapping project – fieldwork in the Khumbu. Coincidentally, we arranged for a three-day workshop in Kathmandu to provide a platform for open discussion of our results with local government and university departments and representatives from ICI-MOD and the several bilateral and UN agency aid and development organizations. It was assumed that this would be an opportunity for our entire team to engage in debate, both among ourselves and with the invited participants. The fieldwork in the Khumbu region had begun in the 1982 post-monsoon season (see below), while the Kathmandu workshop was scheduled to be held at the Shangri-La Hotel from 29 October to 1 November 1983.

The Kakani fieldwork was more or less finished, although we hoped to have subsequent visits for replicate photography and to maintain contact with some of the local farmers who had helped us. This transition period coincided with the end of Walther Manshard's term as UNU vice-rector although he was invited to stay on in Tokyo as senior advisor. The new vice-rector, Dr Miguel Urrutia, a Colombian economist, thus became our "chief-of-staff." Accordingly, I invited him to the Kathmandu workshop.

The first of the Kakani maps had been printed in time for the workshop and hand-coloured versions of the others had been prepared (the time was prior to computers). Hans Kienholz and his Bernese colleagues set up an impressive display that included the maps, large diagrams and photo-enlargements and provided an excellent setting for the first occasion on which we were able to present a strong case for our growing concern about the development paradigm of the day, the theory of Himalayan environmental degradation.

In the midst of the workshop, Dr Urrutia invited me to his hotel room for a personal discussion. After expressing his approval for what we had accomplished, he informed me that he thought we had spent enough time in Asia and our project should move on to other mountain regions, such as the Andes. This was a shock. A rather tense conversation ensued. I did my best to explain that the planned work in the Khumbu had barely begun and that there was already a close understanding with the Chinese Academy of Sciences for a major expedition to northwest Yunnan. I proposed a compromise. I would make a personal exploratory excursion to some part of the Andes the next year (he said it could be of my own choice, but the tropical Andes were to be preferred).[73] The Yunnan project would remain on schedule, and the Khumbu mapping project was to be completed. However, the third phase of our Nepal project, examination and hazard mapping of a transect from the Siwalik range out onto the terai and the Ganges flood plain, would have to be postponed indefinitely (i.e., abandoned!).[74] Urrutia then delivered his final blow: he thought that, after three years of financial support from UNU, our new journal *Mountain Research and Development (MRD)*, should be able to stand on its own, supported entirely by membership subscriptions, page charges, and Western university funding. He found it politically inappropriate for UNU to be subsidizing an "American" publication (actually, the journal was legally a joint IMS-UNU publication). This left me in a wretched position, but at the time I thought the compromise I had negotiated was the best that could be done about a threatening situation and it was prudent to play for time.[75]

Fig. 83. In places the Kakani ridge landscape has been so completely engineered by human hands through the generations that no vestige of 'untouched nature' remains. This is a dry season view of bari terraces. We came to resent such beautiful landscapes being characterized as wilful destruction of the environment by "ignorant" peasants (1980).

Mountain Hazards Mapping amongst the World's Highest Mountains

The first phase of the fieldwork in the Khumbu was completed in the 1982 post-monsoon season. On that occasion, we had attracted Dr Inger-Marie Bjønness to our team. I had met her accidentally in Kathmandu while visiting Dr Ratna Rana. She was a Norwegian geographer with Nepalese experience and a working knowledge of Nepali. She proved eager to join our team. I had also recruited Barbara Brower, a University of California, Berkeley doctoral student with experience in pastoralism. She had the task of studying the impacts of modern tourist trekking on the use and deployment of yaks in the Khumbu. Another addition to the field team was Dr Colin Thorn, University of Illinois, who had completed his PhD with me on the frost and snow bank processes of Niwot Ridge, Colorado. Once again, the mountain hazard mapping designer and supervisor was Hans Kienholz, and the main workforce consisted of the "Swiss Guard," several Bernese graduate students, together with our Nepali members.

The scale of the field area, that is, the range of altitude and precipitous nature of the landscape, was so enormous that the associated scale of the hazard mapping had to be tightly restricted. A tactical decision was not to extend the mapping onto the glaciers nor above the permanent snowline. We were not a mountaineering team and were not involved with the hazards of climbing Mount Everest (an entirely different type of hazard). Thus our work was restricted to the inhabited landscape, essentially the area below the snowline and glaciers, with the exception of glacier lakes that became a major focus. The emphasis was to be on the inter-relations between mountain hazards and the Sherpa population.

Sherpa society had begun to change rapidly following the closure of the frontier with Tibet by the Chinese government. This had cut out the traditional primary life-support function of the Sherpas as carriers of goods between the Plateau and the Indian lowlands (basically, the exchange of salt from Tibet for cereal from India). The opening of Nepal to the outside world in 1950 as trading with Tibet was curtailed provided a timely new activity for these entrepreneurial people. The guiding of mountaineering and trekking groups changed patterns of yak husbandry, and the rapid growth in trekking lodges and tea houses opened new employment opportunities.

The earlier topographical mapping activities of Erwin Schneider had resulted in the so-called Schneider map of 1977 (revised edition) at a scale of 1:50,000 and with a contour interval of 40 metres. This provided us with an essential base map, and we set our altitudinal range for the hazards mapping between 2000 and 6000 m above sea level. The Khumbu presented several new challenges. Much of the mountainous area to be mapped was technically inaccessible, and we had to use available air photographs and field binocular examination (Zimmermann et al. 1986).

In addition, there was a near absence of documentary records of previous hazard events thus depriving us of accurate description of actual catastrophic occurrences prior to our research. This was partly overcome by the work of Inger-Marie Bjønness and Barbara Brower and their interviews with the local people. They depended on Sherpa oral tradition, which was limited to two generations although only accounts of the larger events that occurred in proximity to human habitations and the main trails could be assumed reliable.

The attitude of the Sherpas to catastrophic events was revealing. They were heavily influenced by their religion and traditional beliefs. Many of the mountain peaks were considered sacred: the homes of gods and goddesses. Major catastrophic events were believed to be expressions of anger of the divinities in retaliation for human misdemeanours; the Sherpa response usually involved acts of propitiation. Similarly, where potential hazards were perceived, such as large boulders precariously situated on steep slopes, offerings were made and prayer flags erected around the threatening phenomena. A good example of this was realization that the Khunde hospital, built by Sir Edmund Hillary, had been located on the cone of a former debris flow. Mani-walls and prayer flags were set up at points along the course of the torrent that the local people perceived to be especially dangerous (as no catastrophic event has occurred since, can we assume that the Sherpa approach is most effective?).

Two somewhat surprising conclusions emerged from the Khumbu hazards mapping project and associated fieldwork. First was the prevailing general assumption that the extreme relief of the region indicated a high level of slope instability. This proved not to be the case; the region was much less geomorphologically active than the standard literature would lead one to believe. The reason was at least partly due to the highly resistant underlying granite-gneiss bedrock (Vuichard 1986). Related to this was the fact that the total monsoon precipitation was much less than hitherto assumed largely because most of the field area is situated in the rain shadow of the Konde-Kang Taiga mountain ridge. Also, there is a rapid drop in maximum precipitation amounts with increasing altitude. Once again, fieldwork demonstrated the danger of accepting pre-existing general assumptions. Nevertheless, our opposing conclusions were tentative, and this led to a later addition to the project: a detailed study of geomorphic processes over a nine-month period (Byers 1986, 1987a, 1987b, 2005).

Assessment of this first phase of the Khumbu fieldwork indicated that the single most serious hazard was posed by the expansion of glacial lakes and the danger of their catastrophic outburst (Icelandic: *jökulhlaup*). The 1977 jökulhlaup from the Nare Glacier below Ama Dablam was well documented because of the extent of the downstream damage that had occurred. All the bridges across the Imja Khola and Dudh Kosi for 35 km downstream from the source had been swept away, several lives had been lost, and sections of the main trail to Namche and the Mount Everest base camp had been damaged. There were local reports of at least two, and possibly five other jökulhlaup during the previous 40 years (Müller 1958; Brower 1983). Thus the jökulhlaup phenomenon was identified for future detailed work in the Khumbu and presumably other areas of the High Himalaya. This phenomenon subsequently became a major concern for the entire Himalayan arc and took on the descriptive name of GLOF – glacial lake outburst flood.[76] It will be discussed in more detail below.

The results of the mountain hazard mapping project to this point led us to be increasingly confident that we were correct in challenging previous assumptions about environmental instability in the Himalaya and its causes. The Khumbu fieldwork, added to that on the Kakani ridge, induced us to question more insistently the worldwide assumption that deforestation by "ignorant" mountain farmers was on the brink of causing environmental, social, and political collapse across the entire Himalaya-Ganges region. In the High Himalaya we clearly needed a prolonged period of geomorphic process studies at least throughout an entire monsoon season. A new graduate student, Alton Byers, was very enthusiastic when I suggested this problem for a doctoral dissertation topic (see below).

Once again, Walther Manshard was able to outmanoeuvre our new vice-rector Urrutia to cover the expenses of Alton and Elizabeth Byers, a geohydrologist also with Himalayan experience, to spend nine months in the Khumbu.[77] Colin Thorn agreed to join the field team as advisor.

First Major Test of Khumbu Environmental Alarms

Alton, Elizabeth, and Colin departed for Kathmandu in March 1984. On arrival they were joined by Khadga and Narendra and ferried by Twin Otter to Lukla's precarious STOL airstrip, gateway to the Sagarmatha (Mount Everest) National Park. From there it was a two-day gentle, acclimatizing walk to Namche Bazar along the lower part of the main trekking route to the Mount Everest base camp. I remember a short letter from Colin commenting on his climb up the final steep section of the trail to Namche. Sections of the trail afforded views westward up the Bhote Khosi valley toward Thame and the site of the Austrian Aid Small-Hydel project, well on the way to completion. "I don't expect the hydro-station will be there for long on a site like that," he wrote. I had always tended to think of Colin as something of a sceptic, a very healthy characteristic as we challenged any number of cherished truisms in the Himalaya. He was prescient, or simply an observant geomorphologist. The nearly completed hydro-electric facility was totally annihilated on 4 August the following year by a jökulhlaup bursting out from one of the glaciers a short distance farther upstream.

Alton had made arrangements for housing at Khumjung, a short distance beyond Namche, for the group's nine months in the field. Khancha Lama and Pembra Sherpa joined them as local assistants. Colin's task was to advise on site selection for the soil erosion study plots and the layout of the instrumentation and equipment. This had to be done before the onset of the summer monsoon, so the entire party was immediately involved in a great high-energy push. Colin, his work completed, was able to return home before the rains came and movement across Nepal became unpredictable.

Thirty-six soil erosion plots were set up through a range of altitude from 3440 m to 4412 m above sea level, which enabled the measurement of precise soil erosion losses through a representative variety of vegetation types up to the alpine treeline. Small climatological stations[78] were established adjacent to the soil erosion study plots, and observations were taken at each site throughout the pre-monsoon, the full monsoon, and the early post-monsoon seasons. Six permanent automatic river gauging stations were also installed along the Imja Khola.

More than 50 soil samples and soil pit profiles were obtained; more than 2000 tree cores were taken; and 2500 plants, together with seeds, collected and pressed, representing over 500 different species. Ten lake sediment cores containing charcoal samples were collected for pollen analysis and radiocarbon dating. Many of Erwin Schneider's landscape photographs from the 1950s and 1960s were replicated by Alton. Finally, six Khumjung households allowed their fuelwood consumption to be monitored for three months.

From the analysis of Alton's observations and data, it was possible to disprove several of those major assumptions ("mountain myths") that our programme had set out to investigate. This bold claim was justified because it was the first time that such a large amount of carefully recorded information had been collected where very little had existed previously.

Prior to the fieldwork, there had been numerous predictions that the environment of Sagarmatha National Park (and World Heritage Site) was undergoing serious damage. The causes were cited as an inter-related combination of uncontrolled deforestation, overgrazing, indigenous population growth, and a rapidly increasing influx of mountaineers, trekkers, and their guides and porters and the pressure they exerted on the environment. These claims had been alarming because they emanated from highly respected academics and forestry experts working as advisors to UN and bilateral aid and development agencies, and to the Nepalese government. The accounts of impending disaster were fuelling the news media blitz of the time. For example, the well-known anthropologist, Professor Fürer-Haimendorf (1975: 97–98) wrote:

> ... *forests in the vicinity of the* [Khumbu] *villages have ... been seriously depleted, and particularly near Namche Bazar whole hillslopes which were densely forested in 1957 are now bare of tree growth.* (quoted by Byers 1987b: 210)

while Hinrichsen et al. (1983: 204) claimed:

> ... *more deforestation* [has occurred in the Khumbu] *during the last two decades than in the preceding 200 years ....* (quoted from Byers 1987b: 210)

These statements of imminent disaster, and many more like them, matched similar statements relating to the Himalayan Middle Mountains and impacts on the Ganges Plain, all of which were bundled together as the theory of Himalayan environmental degradation. We had regarded these predictions with a degree of scepticism, partly because they were so apocalyptic and partly because there appeared to be no solid supporting data. The 1979 reconnaissance to Kakani and the Khumbu, and especially the early fieldwork in the Kakani Middle Mountains, reinforced our scepticism. The Khumbu mountain hazards mapping led to even stronger doubt. However, to publicly refute the published statements of well-known scholars and journalists required a strong data base. Alton's research design was set to provide this, at least for one key area of the High Himalaya and was accomplished

successfully. It proved an effective intellectual capstone for the already extensive research both in the Khumbu and the Kakani field sites.

Following the 1984 fieldwork, it became clear that the ongoing claims about both the longer-term and more recent impacts of human activities on the Khumbu environment needed extensive revision. First, the south-facing mountain slopes below the local tree-line had been widely described as unstable and incurring serious soil losses. These slopes were cut by numerous terracettes, caused by the trampling of yaks. Observed in the pre- and post-monsoon seasons (the usual times for visitors, mountaineers, trekkers, or researchers) they appear almost bare, with little vegetation cover. Thus it would seem that, with the onset of the torrential rains of the monsoon, soil losses would be immense. However, for rare monsoon-season visitors, such as the 1984 UNU team, it became evident that light rains preceded the main onset of the monsoon and the apparently scant ground vegetation quickly spread to provide near-total protective cover before the onset of any heavy rain. Furthermore, the monsoon rains themselves proved to be comparatively light. The south-facing slopes, based on the soil erosion plots, experienced minimal erosion; the north-facing slopes, largely under forest cover, even less. The replication of

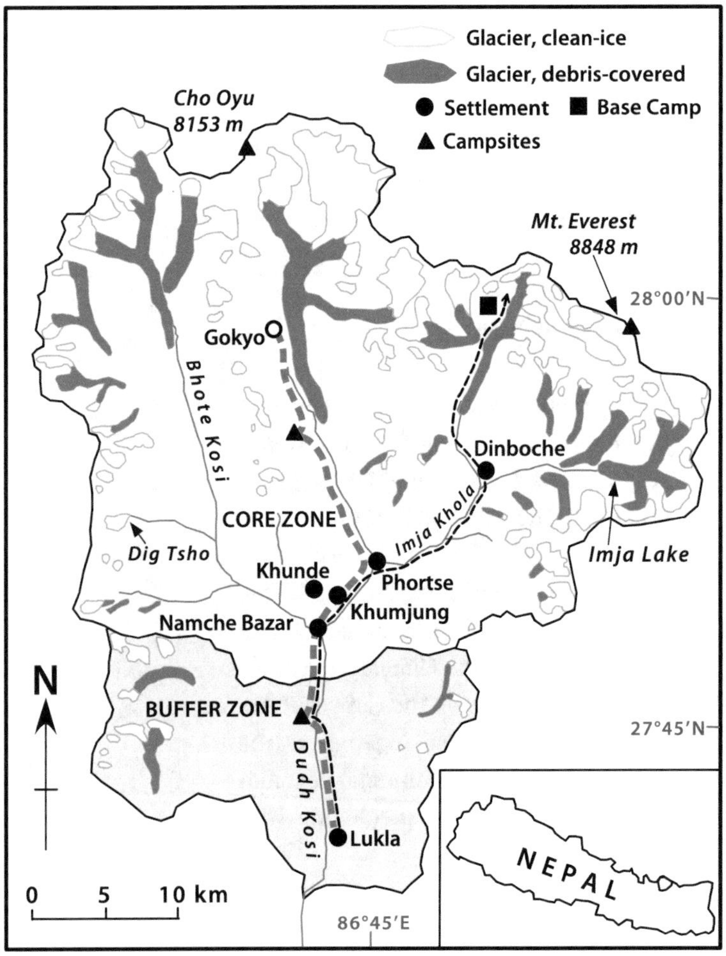

Map 7. Sketch map of the Khumbu Himal. The dotted line indicates the route followed by the 1979 reconnaissance mission to Gokyo via Namche Bazar. The main trekking route to the Mt Everest base camp is also shown together with Imja Lake and Dig Tsho. Repeat of Map 5, page 98.

Schneider's landscape photographs from the 1950s and 1960s indicated little change in forest cover. The hillsides that were claimed as having been stripped of their 1957 forest cover were shown to have had no forest cover on Schneider's photographs, taken before the 1957 claim – an embarrassing exposé.

Similar studies in the Langtang Himal carried out by Teiji Watanabe, another of my doctoral students, produced corroborating results (Watanabe 1994). This was important because it demonstrated that our conclusions were not based on work in a single area.[79]

There was one important exception to Alton's claims of very slight soil loss. At his highest study plots above the treeline in the vicinity of Dingboche and in the juniper-shrub/subalpine belt, he recorded conspicuous damage. Here juniper shrubs were not only being extensively cut, but their root systems were being pulled out, thus laying bare a disturbed soil surface and subjecting it to extensive erosion. The causes were collection of fuelwood for the trekking and mountaineering groups, traditional Sherpa use during the upward movement for summer grazing, and the need for juniper in religious observances. Nonetheless, the National Park authorities were planting seedlings at lower elevations where, although aesthetically desirable, they were hardly needed to replace trees incorrectly reported to have been stripped away.

Another widely accepted folk story was that the first people to immigrate into the Khumbu were the Sherpas who had arrived from Tibet about 400 years earlier. Alton's pollen and charcoal profiles, when radiocarbon-dated, made it certain that the Khumbu had been inhabited for far longer than 400 years, possibly more than a thousand years. It could not be proven whether these were early Sherpa arrivals but it could be hypothesized that animal grazing, accompanied by deforestation had been practiced for much longer than hitherto assumed. This would provide a very long timeframe for the clearing of the south-facing slopes as subsistence grazing expanded and, presumably, was accompanied by logging to support local needs.

The Khumbu hazard mapping component of the UNU project had already concluded that the single most serious danger facing the inhabitants and visitors was the potential for catastrophic drainage of glacial lakes. The destruction of the Thame Small-Hydel Project (with accompanying loss of life and Sherpa property) on 4 August 1985, however, set in motion a long-term involvement that has continued to the present.

Glacial Lake Outburst Floods in the Khumbu and Wider Implications

In August 1985, while I was in Bern, news reached us from Kathmandu of the outburst of a glacial lake (Dig Tsho) from the Langmoche Glacier in western Khumbu. The consequences were disastrous: the almost completed Austrian Aid hydro-electric project situated about seven kilometres below was totally destroyed, several lives were reported as lost, and there was significant property damage for 70 kilometres downstream. On Walther Manshard's confirmation in Tokyo of a supplementary budget item, Daniel

Vuichard and Markus Zimmermann were able to fly immediately from Bern back to Nepal and spend three weeks in the Dudh Koshi and Bhote Koshi valleys conducting an assessment of the damage (Vuichard and Zimmermann 1986, 1987). This early record of the Dig Tsho *jökulhlaup* prompted Dr Colin Rosser, director-general of ICIMOD, to ask me to prepare a detailed assessment.

I spent most of July and August in Kathmandu the following year, and the report was published as *Glacial Lake Outburst Floods and Risk Engineering in the Himalaya* (Ives 1986). It represented the first published study of a Himalayan catastrophic glacial lake outburst and the damage and loss of life that resulted. It recommended an urgent search for similar potentially dangerous lakes in the Nepal Himalaya. It also urged that, as other glacial lakes were located (and we were aware of several), there should be an evaluation of potential downstream impact, identification of endangered infrastructure, and introduction of land-use planning. I was greatly assisted by the reports prepared by Vuichard and Zimmermann. Dr Vic Galay, a Canadian hydrological engineer working as a consultant for the Nepalese Water and Energy Commission Secretariat, who had also been to the site, generously provided access to his records.

During the Khumbu mountain hazards mapping in 1982 and 1983, Hans Kienholz and his team had obtained information on earlier glacial lake outbursts. They found that the most serious was the one draining from the Nare Glacier below Ama Dablam in 1977, although it had not been adequately investigated.

My report put together all the available information and evaluated the original development of the hydro-electricity facility that had been destroyed. I was critical of its actual location and especially of the fact that the level of the glacier lake, Dig Tsho, had been known to have risen almost to the crest of its end moraine dam at least two years before the outburst. Examination of the available satellite imagery augmented the findings of Hans Keinholz's group and demonstrated the existence of several other glacial lakes, in the Khumbu, in the Rowaling Himal to the west, and in the upper Arun drainage to the east on both sides of the Chinese-Nepalese international border. Consequently, I recommended to Colin Rosser that ICIMOD had a timely opportunity to initiate a new Himalayan glaciological research project with emphasis on the danger posed by the glacial lakes. He maintained that ICIMOD was not a research institution and, in any event, he was sure that his Board of Governors would not approve.

One of the largest glacial lakes in the Khumbu was observed on the lower tongue

Photograph on facing page
Fig. 84.   Aerial view of the glacial lake, Dig Tsho, located at the snout of the Langmoche Glacier in the western Khumbu Himal. The lake breached its damming end moraine on 4th August 1985 causing extensive damage for more than 70 km downstream. This photograph, taken on 24th April, 2009, demonstrates that the cut through the end moraine dam in 1985 was so complete that only a very small lake has reformed, posing little danger. (Photograph courtesy of ICIMOD, taken by Sharad P. Joshi).

of the Imja Glacier some ten kilometres south of the summit of Mt Everest. I later realized that the Imja Glacier and its lake had been photographed from the air several times in the 1970s by Japanese colleagues although in 1986 it had no generally accepted name and its development history and risk of outburst had not been studied (see Vuichard and Zimmermann 1987).

Later, at home in Colorado, I examined the contents of a large box of photographs that had been taken by Fritz Müller during his participation in the 1956 Swiss expedition to Everest and Lhotse. Fritz was a colleague from my graduate student days at McGill University, Montreal, who had died tragically in Switzerland in 1980. His former student, Konrad Steffen, brought the box to me when he moved from Zurich to Boulder. It had remained only briefly examined, but now I was prompted to investigate.

To my great satisfaction, there were more than a dozen high-quality enlargements, including two panoramas of the lower Imja Glacier. They showed, without a doubt, that no glacial lake existed in the 1950s and only a few small melt ponds could be detected amid the jumble of surface moraine.[80] The next step in what was to become an exciting hunt for potentially dangerous glacial lakes in the Himalaya was a telephone conversation with my old friend Barry Bishop.

Barry had flown with Dr Bradford Washburn in December 1984 to direct the air photography for the National Geographic Society's topographical map of the Everest region. He outlined my interest to Dr Washburn who promptly sent me a stereographic pair of air photographs of Imja Glacier and its surroundings. When compared with the ground photographs from 1956, it was apparent that a large lake had formed during the following 28 years. Its area in December 1984 was about 0.6 km² and appeared to be enlarging.

Thus began a search for all available information on the formation and growth of what eventually became known as Imja Lake, or Tsho Imja (*tsho* is dialect for lake). June Hammond, one of my graduate students, used the air photographs and research reports as the basis for her master's thesis (Hammond 1988).

June obtained assistance from several of my colleagues, especially Professor Keiji Higuchi. This included a series of photographs taken from the air during the Japanese glaciological expeditions to the Khumbu in the 1970s. Her work was carried forward by Teiji Watanabe, who has spent many subsequent field seasons in the Khumbu and undertaken increasingly detailed surveys of the Imja Glacier and Imja Lake, eventually working in collaboration with Alton Byers. Their research, and that of Teiji's graduate students, have continued up to the time of this writing (Lamsal et al. 2011; pers. comm. Byers 2012; Watanabe 2012).

During the early phase of our interest in Imja Lake, its rapid and continuing expansion led to concern about the possibility that it could burst out catastrophically. Such an event would threaten several villages in the valley below and destroy long sections of the main trekking route to the Mount Everest base camp and to Namche Bazar. If an outburst occurred during the trekking season, there

170

Fig. 85a.   The Imja Glacier (extreme right) in 1956, Khumbu Himal. At this time no supra-glacial lake has formed. The lower glacier is mantled by morainic debris and rockfall. There are several small melt-ponds between the rubble. (Photograph from the collection of the late Fritz Müller)

Fig. 85b.   By 2007 the lower part of the Imja Glacier is occupied by a lake (Imja Lake) more than two km long and as much as 100 m deep. The glacier is still retreating and the remaining sub-lacustrine ice is continuing to melt. The view is from the opposite side of the glacier from that of Fig. 85a. The spectacular Mt Ama Dablam forms the right hand horizon. (Photograph courtesy of Dr Alton Byers)

Fig. 86. Aerial view of the Imja Glacier showing Imja Lake, 4 November, 1991. At this point in its rapid development the lake is more than a kilometre in length. It drains through the broad end moraine in the immediate foreground, the source of the Imja Khola (Photograph courtesy of K. Kawaguchi).

would be a much higher loss of life and probable international repercussions. I suspected that, at the time, His Majesty's Government of Nepal did not want such an alarm to be publicized as it would likely curtail the number of tourist visitors and so reduce revenue (a rather cynical proposition!). A major problem (1986–1992) was that we had no way to determine either the volume of the accumulating melt-water or the sub-surface condition of the end moraine dam, so that prediction of hazard was highly problematic.

The opposition of Vice-Rector Urrutia to continued UNU funding for fieldwork in the Himalaya after 1983 meant that work on the Imja Glacier was severely restricted. June Hammond's study was virtually unfunded, and Teiji Watanabe had to rely on limited support from standard Japanese sources. However, my main source of frustration was the unwillingness of ICIMOD Director-General Colin Rosser to see the potential and also the possibility for damage to property and the loss of life of people living downstream. The major problem was that there was totally inadequate information available for any attempt to determine the degree of the hazard.

This discussion of the possibility of jökulhlaups in the Nepal Himalaya has been given in detail because it will lead into the second major alarmist theme of this book that is taken up in Chapter 16.

Fig. 87. The high peaks of the Yulong Xueshan (Jade Dragon Snow Mountains: 5596 m) support the most southerly of Eurasia's glaciers. The small village of Yuhu can be seen in the bottom left, home of Dr Joseph Rock for much of the period between the mid-1920s and 1947.

CHAPTER 11

China Revisited 1985: Yunnan and the Jade Dragon Snow Mountains

Until 1985, most of our UNU mountain project resources had been deployed in the Nepal Himalaya (Chapter 10), although the contact with Sun Honglie and the Chinese Academy of Sciences was seen as extremely important and led to several years of field research in Yunnan. As described in Chapter 9, we had undertaken reconnaissance expeditions to the Tian Shan in 1981, and to the Hengduan Mountains in western Sichuan and northwestern Yunnan in 1982. The Lijiang-Yulong Xueshan area of Yunnan appeared the most suitable. Access was relatively easy and, in particular, several critical field problems that called for attention could be investigated in quite close proximity. One of the most appealing aspects of this area was that it had been extensively photographed by Joseph Rock throughout the period 1923–1947, and the majority of his photographs had been archived by the National Geographic Society. We had been promised access to these by my friend Barry Bishop who, at the time, was vice-president of the society and chair of the Committee on Exploration and Research. Furthermore, Barry was keen to accompany us to Lijiang.

I recommended to Honglie that we investigate the current rate of deforestation and its causes, focusing on the chronology and manner of forest loss since 1950. With access to the Rock photographs, we should be able to determine any forest cover changes throughout the period of his sojourn and compare them with the current situation if we could locate his actual photo stations in the very rugged landscape. Bruno and I had already tested such an approach by replicating some of Imhof's photographs and water-colours from his 1930 expedition to western Sichuan (Gongga Shan, or Minya Konka). To attempt something similar, but on a larger scale in northwestern Yunnan would be a priority. I suspected that this approach would lead to controversial findings similar to those emerging from our research in the Nepal Himalaya. In other words, we would be able to test our hypothesis that the extensive deforestation throughout Yunnan was not solely a recent (post-1950) phenomenon but had occurred over a much longer period of time.

A second objective would be a preliminary study of the current way of life of the Naxi and Yi peoples and of their relationship with the Han, especially in terms of the use of natural resources and the sustainability of their land-use systems. These questions, of course, would underpin the issue of mountain hazards and soil erosion.

At the time the draft proposal was being prepared, most of China was still closed territory, and we were hoping for access to mountain minorities, most of whom had never seen foreigners. The proposed field area had not been visited by foreigners since before World War II and by very few before that. Much of Yunnan was dominated in the 1920s and 1930s by warlords, opium production, and extensive conflict. Joseph Rock always travelled with a heavily armed guard of Naxi warriors. Consequently, our plans needed approval, not only by CAS and the Beijing central authorities, but also by the People's Liberation Army and the Governor of Yunnan, as well as the local villagers. Given our near disaster in offending the local Bai people in September 1982 (Chapter 9) these hurdles were rather intimidating. However, our proposal was endorsed promptly by all levels of administration, placing us in a highly privileged position.

The draft proposal to Honglie was approved with minor changes; like our proposals for the previous reconnaissance expeditions, it was translated into Chinese and submitted to UNU in Tokyo. The plan called for ten weeks in the field, two Western researchers and two graduate students, with CAS supplying four senior scientists, two junior assistants, and support personnel, including drivers with their vehicles, Naxi and Yi interpreters, and a cook. All the food would be acquired locally, and the CAS personnel would supply their own field equipment. I took a minimum of lightweight mountaineering and camping equipment and a small supply of Kendal Mint Cake, a last resort for Western visitors in a near-sugarless environment. As it turned out, we should have brought much more equipment with us, even hiking boots and lightweight water flasks for our Chinese colleagues – this still was a period of enormous contrast between China and the West in development of camping equipment. However, CAS established an excellent base camp – a renovated small World War II American airstrip and buildings located about 20 km north of Lijiang.[81] Honglie sent a work crew to renovate the buildings and install a fully equipped kitchen. Two young members of the People's Liberation Army accompanied us. I was never sure whether their duty was to protect us from the local people – the very kindly and friendly Naxi villagers – or them from us![82]

Late in the planning stage, Barry Bishop had to withdraw due to internal National Geographic Society politics. His request for ten weeks leave may have exceeded reasonable limits. However, as a vital contribution, Barry put together a representative group of Rock's photographs, beautifully enlarged and printed – they arrived on my home doorstep the day before I set out for Beijing. These photographs provided the basis for one of the more significant findings of the project. As Barry was unable to join us, I decided to take only one Boulder graduate student with me, Alton Byers. In Alton I obtained a seasoned field colleague, fresh from his highly relevant research experience in the Khumbu Himal.

Yulong Xueshan: The Jade Dragon Snow Mountains

The Yulong Xueshan region is transected by the latitude 27°N and longitude 100°E. It lies in the great double bend of the Jinsha Jiang (upper Yangtze River) about 650 km northwest of Kunming, the capital of Yunnan Province. It forms the core of the Naxi Autonomous District with a population of about 300 000. The Naxi people are the majority, with smaller numbers of Han, Yi, Bai, and Tibetans. The regional administrative centre is Lijiang, traditionally known as Dayan, the 800-year-old capital of the Naxi; the population is about 60 000 and the town lies at an altitude of 2475 m. The highest summit of the Yulong Xueshan is Shanzidou, 5596 m, unclimbed in 1985. It towers above Lijiang some 20 km to the north and is the culmination of an impressive range of limestone peaks. Due west of Lijiang, the Jinsha Jiang makes its first impressive hairpin bend to flow northward between the Yulong Xueshan and the Haba Xueshan (5396 m), forming one of the world's most spectacular river gorges (*A cai ggoq*, or Tiger Leap Gorge, so named from Naxi folklore).[83] The gorge accentuates a local relief of over 3700 m within a horizontal distance of less than 6 km between the highest peaks on either side. This change in altitude at latitude 27°N gives rise to an enormous range of vegetation types, from the uppermost limit of banana through multiple forest belts and across alpine tundra to Eurasia's most southerly glaciers.

The Lijiang region's vegetation and wildlife is as impressive as its topography. We were informed that, until recently, the region had supported tigers, leopards, bears, red pandas, wolves, deer, and a large number of small mammal species. We quickly came to realize that much more of the wildlife persisted than this report implied. For instance, I was offered pelts of both the snow leopard and the clouded leopard for a few dollars. We were also informed of the presence of the leopard cat, a smaller feline. More isolated villagers told of depletion of their livestock by leopards and wolves. The birdlife was equally diverse, particularly the range of pheasant species and birds of prey.

Before I left Boulder for Beijing, an acquaintance who was a senior officer of the World Pheasant Association, became very excited when he learned of my destination. He had concluded that an important local species, *Crossoptilon crossoptilon lichiangense* (White-Eared Pheasant), had become extinct. Nevertheless, he provided me with a colour plate to assist with identification in the event of any possible sighting and pleaded that I keep my eyes skinned. While I failed to locate a living specimen, sadly, I was able to take him a photograph of a *Crossoptilon* pelt that was on sale in the Lijiang market place (for less than US$1) to prove that the species was not extinct. Among the rich variety of plant life were more than 30 species of rhododendron, both arboreal and shrub which in May and June set the landscape ablaze.

Field Investigations in the Lijiang Area

The logistics constraining our programme proved more intransigent than

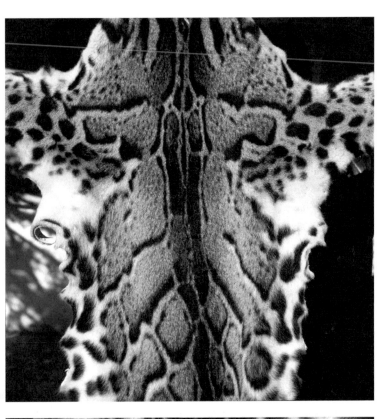

anticipated. The topography was extremely rugged, and sections of forest were almost impenetrable. And the customary Chinese approach to fieldwork, very different from all our previous experience, proved a challenge. For instance, as a team we had only our own two lightweight water flasks. In a region largely underlain by limestone, surface water was scant, and we were expected to rely on bottles of beer and the large and fragile Chinese thermos bottles for tea (neither of which were appealing thirst-quenchers for Alton or myself). The smooth-soled shoes and boots of our hosts were unstable on steep slopes and useless on snow. Sleeping bags and tents were of the size and weight that required a truck for transport, or pack animals when off-road. We were also taken aback to learn that we were expected to return to base camp for cooked lunches. The Colorado contingent regarded a return to camp for a midday meal as out of the question. In like manner, the Chinese were reluctant about setting out from camp before sunrise. Many adjustments were necessary due to the great differences in ethnicity and custom, but the great spirit of camaraderie ensured excellent working relationships.

Chen Chuanyou had been appointed "administrative" leader; I had been accorded the role of "scientific" leader. It proved a good partnership and we became firm friends. He ensured that his Han colleagues

Fig. 88 [top].   Pelt of clouded leopard offered for $15 in a local market.

Fig. 89 [bottom].   Pelt of unidentified species of leopard offered for sale under similar circumstances.

accommodated my "intemperate" approach to fieldwork, and I was happy to welcome him for a three-month stay with me in Colorado the following winter.

The size of our chosen field area and the complexity of its geography (both physical and human) led us to realize that our original plans were too ambitious and would have to be adjusted. The revised plan entailed a reconnaissance of the entire area with three or four specific objectives that could yield well-defined and useful results. It included a soil erosion transect, replication of Rock's photographs, vegetation transects from the lowest levels to as high as we could go, and assessment of changes in forest cover over as long a time period as possible. In the process of these activities, we would learn as much as possible of the ways of life of the local people.

Alton was able to apply his experience by replicating the soil erosion study plots that had functioned so well the previous year in the Khumbu Himal. He trained the two junior members of the team to maintain measurements and operate our weather station through the coming monsoon season after the main group had left for home.[84]

I focused on replicating Rock's photographs and, with the aid of our interpreters, interviewing village elders, particularly on their recollections of the condition of surrounding forest cover throughout their lives. All the team at some time assisted with these efforts, especially Alton, who took cores from more than a thousand trees. These samples enabled us to date the patterns of forest growth and deforestation and gave us a fairly precise chronological

Fig. 90.   The market in Lijiang frequently had the magnificent plumage of the endangered Lijiang pheasant for sale (*Crossoptilon crossoptilon lichiangense*). We were informed that the Peking Opera was a primary customer.

control for the replicate photography.[85]

We soon, however, encountered a serious obstacle with the photography. The 1985 pre-monsoon season was unusually humid. We had expected that the region would have been under the influence of a northwest air mass moving gently off the arid high plateaus of Central Asia. Instead, clouds began to form on the mountains within one or two

Fig. 91.   In 1985, this young Yi mother with her baby was bribed to pose for the photograph by our promise of a 40 km ride in our van into Lijiang, together with her family and market produce – a long walk!

hours of sunrise and did not dissipate until late afternoon when the eastern flank of the Yulong Xueshan, our main camera target, was in shadow. This meant that we had to locate Rock's photo stations and then return on a subsequent day to take photographs early in the morning before clouds obscured the view. On several occasions, because the photo stations were so far from the camp, we had to set out in the dark.

The Yi driver of our ten-seat van, the most versatile vehicle in our fleet, was promised extra pay if he would exchange his hot breakfast for cookies and a thermos of tea so that we could set out before dawn. However, when he understood our needs, he refused

the extra pay; he came to regard the early morning drives as a special kind of holiday. Some photo stations proved to be inextricably lost in dense forest that had sprung up since Rock had been on the site and could not be located. While this deprived us of several replications, it also proved that, in some areas, forest cover was much more complete than at the time of Rock's visits in the 1920s and 1930s.

A journey was made to one of Rock's favourite Naxi villages. Wenhai is situated on a scenic lake[86] at 3110 m above sea level at the southwestern foot of Shanzidou, highest peak of the Yulong Xueshan. The village is a day's march from the nearest road-head to the south. Chen Chuanyou hired horses

Fig. 92.   Our welcome to Wenhai in 1985 was accompanied by an offer to guide us to the location where Rock had taken his photographs in 1928. The observant Party Secretary, second from right, also demonstrated his political astuteness. Dr Zhang Yongzu is on his right with camera (see text page 182).

Fig. 93. Dr Joseph Rock took this photograph in 1928 when he visited the village of Wenhai. The time is immediately post-monsoon so that the lake level is at its maximum (Courtesy, Dr Barry C. Bishop, National Geographic Society).

and handlers, and the entire team, except the cook, set off for a four-day excursion.

The village elders were well aware that Rock had visited their village several times during his Yunnan sojourn. When the local Communist Party secretary was shown the two Rock photographs that I had selected for replication, he immediately asserted that he could find the places where Rock had set up his tripod. He walked me through the forest and placed my camera tripod in exactly the same spot that Rock's had been all those years ago. He also made a surprise request; would I send him copies of both photo-pairs? He was well aware that the forest landscape of 1985 was in much better condition than when Rock had taken his photographs. He was up for re-election in 1986, and his election platform included opposition to a road planned by the district government. He insisted that to provide improved access to this isolated village would lay open its precious forests to illegal logging under cover of darkness by villagers from farther south. The photo-pairs would allow him to demonstrate that his policy of forest protection had produced sound results. One of the photo-pairs, reproduced as Figures 93 and 94, became an iconic item in our attempts to demolish the Chinese version of the theory of Himalayan environmental degradation.

Fig. 94. Replicate of Rock's photograph taken in April, 1985. As with the following photo-pair, the April view shows that the lake has drained almost completely into the underlying limestone. This view shows that the forest cover is much more complete than 58 years previously (1985).

Another interesting aspect of the photo comparison at Wenhai is that Rock's photographs, taken post-monsoon, show a sizeable lake. The ones taken in April 1985, pre-monsoon, have no lake. The lake basin, underlain by limestone, contains one or more dolines (sink holes, features of karst topography). During the dry season after the monsoon, the lake water progressively drains away and refills during the following monsoon.

We also made a short expedition to the mystical sub-alpine meadow I had briefly visited in 1982 – the so-called Love-Suicide Meadow. We camped there for three nights close to the very spot where Rock had pitched his tents. Our view was essentially identical to his. Alton cored several dozen trees and calculated that the older trees were more than 900 years old. Sustained grazing by domestic animals over the centuries had preserved the meadow from invasion by seedlings and shrubs. This was yet another sign of the long-term stability of at least some areas of northwestern Yunnan.

On our homeward journey, I replicated one of Rock's photographs of the old city of Lijiang. His photo station was easily located.

Fig. 95.   Photograph taken by Joseph Rock in 1928 showing the east face of the Yulong Xueshan about 40 km north of Lijiang. Rock's camp can be picked out on the far side of the lake (Courtesy, Dr Barry C. Bishop, National Geographic Society).

Fig. 96.   Replication of previous photograph taken in 1928. The boulder in the foreground confirms the accuracy of the photo station. April, towards the end of the winter dry season has witnessed the progressive drainage of the lake into the underlying limestone bedrock. While the forest cover in the middle ground appears similar to that on Rock's photograph it is practically all post-1928 growth as determined by tree-ring count (1985).

The two shots, taken almost 50 years apart, provided such a precise comparison that many of the same individual rooftop tiles could be identified; in other words, little architectural change could be detected.

Summing Up, 1985

The Chinese equivalent of the theory of Himalayan environmental degradation had been refuted. This, in turn, greatly strengthened the conclusions emerging from our work in Nepal. The comparable results from the two widely separated regions could be combined for presentation at the Mohonk conference the following year (Chapter 12). We were able to show that, in some areas of northwestern Yunnan, the forest cover was more complete in 1985 than a half-century earlier. In other areas the situation appeared balanced. In yet other areas the degree of clear-cutting shocked us – not only the cutting, but the extensive wastage, as great numbers of logs were left to rot due to inadequate transport facilities. Nevertheless, there was no longer any

Fig. 97 [top]. Management of the local forests is emphasized by this 1985 photograph. A new crop of saplings is evident following early extensive cutting with the occasional mature trees being preserved as a seed source

Fig. 98 [bottom]. Dr Joseph Rock's camp in the 'Love-Suicide Meadow' in 1931. The highest summits of the Yulong Xueshan (Shanzidou, 5596m) form the centre horizon (Courtesy Dr Barry C. Bishop, National Geographic Society).

doubt that claims of massive deforestation across the entire Yunnan Plateau and in much of the Hengduan Mountains since 1950 were totally insupportable. Certainly, deforestation had occurred since 1950 and was still occurring, but on nothing like the scale necessary to support the prevailing claims. As in other regions, the doomsday conclusions of the day were based largely on unsupported assumptions rather than solid fact. Subsequently the Nature Conservancy undertook an extensive study involving photo-replication that 'in general' further substantiated our conclusions (Moseley 2006).

What we had learned of the lives of the local minority people, of the early deforestation of large parts of the Yunnan Plateau, of the current complexity of forest and general resource management, was sufficient to justify future visits and more intensive investigation.

As a departing gesture, the Lijiang district governor asked if I would give a public lecture (I had brought with me some 35mm slides on the off-chance that even in remote Lijiang a projector could be found, and this proved to be the case). So, assisted by an interpreter, I talked about mountains and environmental conservation emphasizing the UNU work in Nepal to an audience of townspeople and elders from the outlying villages. In thanking me, the governor explained it was the first public lecture ever given in Lijiang by a foreign guest and he had never seen colour slides before. He graciously presented me with the

Fig. 99.   The same view as the previous photograph but taken in April, 1985. Our camp is on the opposite side of the meadow from Rock's camp of 1931. The high summits are in cloud. The tall trees on the far side of the meadow were dated at 500 to 1,000 years old. Alton Byers, in broad-brimmed hat, is second from left, Dr Zhang Yongzu is in the centre.

keys to the city,[87] and as a tangible memento he gave me a traditional set of Naxi lock and key which has a place of pride on my desk to this day.

Fig. 100.   High in the Swiss Alps from which sprung the inspiration for Mountain Agenda and where "the road to Rio" (UNCED 1992) began.

CHAPTER 12

The Mohonk Process
Mobilization of the Mountain Advocates

The "mountain journey" actually followed several inter-related pathways, some of which overlapped in time, and all became mutually reinforcing. The two international mountain conferences held in 1982 and 1986 at Mohonk Mountain House, New Paltz, NY had so many ramifications that this chapter is devoted to them and to what later became known as The Mohonk Process (Thompson 1995; Forsyth 1996). Its very origins derived from a remarkable piece of serendipity, although such happenstance has been by no means unique in this mountain journey.

The Mohonk Mountain Conference of April 1986 (titled "The Himalaya-Ganges Problem") was one of the major threads in the endeavour to alert the world-at-large to the importance of mountains and mountain peoples. Because of the extensive influence of the Mohonk Process I will introduce it through a highly personal experience that was the spark that set it in motion.

Mohonk Mountain House is a renowned five-star hotel situated on the edge of the Shawingunk Mountains in upper New York State. The original buildings are wooden and Victorian in both age and beauty; the hotel

was founded by the Smiley family, well-to-do Quakers, in the 1880s. It was subsequently expanded so that, by the 1980s, the hotel had become a highly regarded recreation site for the well-heeled of New York City. The old structure and its more modern additions stand on the margin of a scenic, partially rock-bound lake in very extensive grounds. By the time of my first visit, most of the estate had been bequeathed to New York State as a biological protected area and state park. The Shawingunk escarpment had become the third most popular rock-climbing locale in the United States after Yosemite and the Colorado canyons around Boulder. Subsequently, management of the hotel itself was transferred to a commercial enterprise although the Smiley family retained space as the headquarters of its Mohonk Preserve foundation, and several of the family continued to occupy private homes in the general vicinity.

The Mohonk Preserve's primary mission was to encourage natural science research within the state park by awarding grants to scholars and students in the various life and earth sciences. By the late 1970s, several members of the Preserve's board of trustees, all

J. D. Ives, *Sustainable Mountain Development*, https://doi.org/10.1007/978-3-030-96029-2_12

family members, had begun to sense a need for broadening its mission. Therein lay the springboard for a series of fortuitous events that provided the basis for the next major breakthrough crucial to eventual worldwide mountain awareness.

Bradley Snyder, an ardent rock climber and doctoral candidate in German literature, had chosen the University of Colorado, Boulder, for his graduate school, in part because of the proximity of first-class climbing rocks. Upon completion of his doctorate, and without being aware of our presence in Boulder, he became executive director of the Mohonk Preserve. The prospect for climbs on the Shawingunks must have been a considerable secondary attraction.

Despite his transfer to New Paltz, NY, and Mohonk Mountain House, he continued to subscribe to Boulder's local newspaper *The Boulder Camera.* In one issue in the fall of 1980, he read of the founding of a new society – The International Mountain Society (IMS) – with myself as president, Roger Barry as secretary, and Mischa Plam as treasurer. Without prior reference to us, he had suggested to the trustees of the Mohonk Preserve that this new society might be a source for fulfilment of at least some of the members' aspirations to expand the Preserve's mission – perhaps an international conference on problems facing the world's mountains?

So Brad Snyder came to Boulder in August 1981, to meet and interview the three officers of the IMS and to collect ideas. He instantly appealed to us as a very attractive and concerned personality. During the course of several long discussions, the notion was floated that an international conference on problems facing mountains and mountain peoples in underdeveloped regions such as the Himalaya and Andes might prove attractive to his board. Brad's report to the board resulted in a request that I submit a specific proposal, together with my curriculum vitae. This in turn led to an invitation to spend several days as guest at the hotel for both informal and formal interviews.

One Friday afternoon in October 1981, Brad met me at Kennedy International Airport and drove me through heavy rain to New Paltz and Mohonk Mountain House. The next day, I was shown around this impressive Old World hotel and the widely known Shawingunk crags where Brad spent many off-duty hours. Later in the afternoon I met with individual board members, joined most of them for dinner, with the evening reserved for a formal "interrogation" by the full board, that is, all except Daniel Smiley, the senior brother, who I quickly learned was a majority of one!

It transpired that the board members were not seeking an exacting interview as they had been favourably impressed by my written proposal and the informal meetings. Instead, we spent a couple of hours in general conversation about Third World mountain problems followed by efforts to prepare me for what they were convinced would be the very tough interview in store for the Sunday afternoon. Elder brother Dan Smiley, had been detained in New York City and could see me only on the following day. Dan, I was told, was quite firmly opposed, and I was warned that I would have a difficult task of persuading him to change his mind. They all supported my proposal but recognized that the final decision would be his. They suggested that I should emphasise my new position as coordinator of the United Nations University mountain project, although doubt remained.

At 2:45 p.m. on the following wet Sunday afternoon, Brad escorted me to Mr Daniel Smiley's luxurious home, a few minutes walk from the hotel. At the door, I was greeted formally. Brad was thanked and sent on his way; I was escorted into a palatial study. Two walls were lined from floor to ceiling with expensively bound books. Expansive windows gave out onto a manicured park landscape while a large and gleaming walnut-topped desk occupied a commanding place near the centre of the room. Before the desk was a commodious winged armchair where I was invited to sit. I sank into its luxury as Mr Smiley flipped through a pile of papers on his desk – my proposal and curriculum vitae. I was decidedly tense; perhaps the friendly warnings of the previous evening were producing a negative reaction.

Mr Smiley appeared to be studying one particular page. Presently he looked up: "I see you were married in 1954, emigrated to Canada, and with your new wife began your research studies in Labrador. How did you come to choose Labrador?"

I was completely taken aback. Never before had I been asked the question in such a way. I paused, nervous. Then my answer came tumbling out of nowhere. I had not been conscious of there being a specific reason previously. Mr Smiley had clearly plumbed a hitherto subconscious depth. I stumbled to explain that, as a boy, I had been dyslexic – a "duffer," as I was often called in those far off days. The only school prize that I had been capable of winning was a Church of England Sunday School prize (at about age eight) for being "never absent, never late." The prize was a book: *Storms on the Labrador*. It was about Sir Wilfred Grenfell's work on the Labrador. I had been so impressed by the descriptions of the Labrador coast and the work of the Grenfell and Moravian missions presented by my Sunday School teacher that I had resolved to go there, if ever I had a chance, and Sir Wilfred became one of my heroes.

Mr Smiley looked straight at me, smiled, and said, "Well, the interview is over. You have the grant. The chair you are sitting in was Sir Wilfred's favourite. He used to come here for rest and recuperation after exhausting months on the Labrador. He would set me, as a small boy, on his knee and read stories to me." I was stunned.

As I recovered from the shock, we talked over coffee and cakes about Labrador, Sir Wilfred,[88] and the forthcoming mountain conference. When it came to leaving the house, Mr Smiley directed me to the west wing of the old part of the hotel where I would find a large oil painting of my hero.

On my return to the hotel, an expectant Brad ushered me into the board room so that I could relate to the waiting group how I had fared. When I recited the story to them, they were as startled as me; they were also enthusiastic about the outcome, but especially amazed by the coincidence of my indirect boyhood acquaintance with Sir Wilfred.

The Mohonk Mountain Conferences

The first Mohonk Mountain conference was held the following year, 5–10 December 1982. It was very encouraging that I was able to attract senior representatives of United Nations, including UNESCO, UNDP, FAO, the World Bank, and aid and development and research organizations in Bhutan, Canada, China, Ethiopia, India, Nepal, Switzerland, and the United States. There were representatives from the Appalachian Mountain Club, the National Audubon Society, and the National Geographic Society. Then there were the core members of the IGU/UNU/MAB-6/IMS. John E. Fobes, former deputy director-general of UNESCO, who had played an important role during the 1974 Munich mountain conference, agreed to serve as chairman. The contact with the Canadian International Development Research Centre (IDRC) was to prove far more important than I could have imagined at the time as it provided the first tentative link with Maurice Strong.

The conference was largely exploratory, and a critical base was established that linked the efforts of the new International Mountain Society with UN and bilateral aid agencies and influential individuals. It also set the stage for the much more substantial Mohonk II conference, held in 1986, for which Maurice Strong agreed to act as honorary chairman.

Mohonk II, a natural follow-up on the first meeting, was much more tightly defined with a number of specific and quite ambitious objectives. In particular, it proved the occasion for our first major international challenge to the theory of Himalayan environmental degradation. By 1985, when the preparations for Mohonk II began in earnest, the UNU mountain research projects in the Nepal Himalaya, northern Thailand, and in Yunnan, China, were already acquiring a large database from a series of direct field experiences that enabled us to make this debate the centre point of the conference. We were prepared to argue that much of the worldwide claim that the Himalaya were in dire distress was based more on political expediency than fact, and our research results should produce a radical reassessment for the entire region. The World Bank, for instance, had published a statement in 1979 to the effect that, at the current rate of deforestation in Nepal, no accessible forest would remain by 2000. It seemed logical, therefore, to take as the conference title "The Himalaya-Ganges Problem." Before planning for the conference was beyond its earliest stage, however, several fortuitous events unfolded.

Already in September 1981, prior to Mohonk I, a formative workshop had been held in Switzerland. This was organized by Bruno Messerli and his colleagues under the heading "The Stability and Instability of Mountain Environments." It was held in Bern and at the Aletsch Centre for Protection of Nature, on Riederalp in Canton Valais – a combined meeting of the IGU Mountain Commission, UNU, and UNESCO-MAB. The main theme was an interdisciplinary discussion about definitions of mountain stability and instability with examples from a cross-section of contrasting mountain regions. The proceedings were published in successive issues of *MRD*.

Yet another piece of remarkable serendipity was my reading of a short paper in the winter 1983 issue of the in-house magazine of the International Institute for Applied Systems Analysis (IIASA). It had been written by Michael Thompson, and its title intrigued me: "Why Climb Everest? A Critique of Risk Assessment." The paper fascinated me as it was written by a Himalayan mountaineer and anthropologist whose own scepticism closely paralleled our own. It was a theoretical comparison between the high-altitude mountaineer and the Buddhist traders (the Sherpas), both depending to a very large extent on taking risks, and the sedentary Hindu farmers (the cautious cultivators) whose risk-avoidance strategies were much more akin to the rest of humanity. I wrote to the magazine editor requesting permission to republish it as a "mountain chronicle" in *MRD*. This led to direct contact with Michael, who had a consultancy with IIASA. Thus, Michael became a member of IMS and submitted three manuscripts for publication in *MRD*. The first carried the title "Uncertainty on a Himalayan Scale" (Thompson and Warburton 1985).

Michael served as one of the Mohonk II coordinators and became a long-term colleague. His thought-provoking ideas proved an invaluable infusion at a time when we questioned every institutional "conclusion" about mountain environment and the rural people who were being held responsible for the assumed devastation. One of my favourite quotes from Michael's writing has motivated me since the day I first read it in the initial manuscript he submitted: aimed at the big aid and development agencies – "Policy issues can be approached in two ways. You can ask 'What are the facts?' and you can ask, 'What would you like the facts to be?'"

Maurice Strong and Proposal for a Canadian Mountain Institute

Yet another unplanned incident occurred prior to the second Mohonk conference. Shortly after my arrival from Colorado at the house of Bruno and Beatrice Messerli in Zimmerwald, Switzerland, in the early summer of 1983, the telephone rang. Bruno answered and passed the receiver to me saying it was from Vancouver, Canada. A woman's voice identified herself, with apologies for a Sunday afternoon call, as the secretary of Maurice Strong. She had tracked me down to Bern because she had an urgent request: would I please compose a dozen or so pages of comments on mountain problems for Maurice Strong, who would like to use them in preparation for his VIP address on the occasion of the inauguration ceremonies of ICI-MOD the following December in Kathmandu? It would be appreciated if I could get them to her within ten days! This request, totally unexpected from a very prominent figure, surprised me. I wondered how he could have known of my existence! Over the next couple of days, with assistance from Bruno, the task was completed and the document sent by courier to Vancouver.

Both Bruno and I, along with Sun Honglie, Corneille Jest, and several other members of the mountain fraternity, were invited to the ICIMOD inauguration ceremony in Kathmandu, where we met its newly appointed first director-general, Dr Colin Rosser. To combat the possible effects of jetlag, I arrived in Kathmandu two days early. Hannah Strong, the wife of Maurice, was also early and invited me to have dinner with her. Two days later, I was asked to join both Hannah and Maurice for another private dinner. I sensed that I was being skilfully probed to see if I could be of value in Maurice's future plans to include mountain issues as a prominent topic in what amounted to his worldwide environmental activism.

Maurice, who was also chairman of the board of governors of UNU, asked me about my personal long-range interests and intentions for the UNU mountain project and the International Mountain Society and its journal. The tone of the discussion was very relaxed, even convivial as if we were old friends. There was no intimation that Maurice had any specific plan: just a way of pleasantly spending a free evening as a thank-you for what I had written for him earlier. I must admit, however, that a senior contact in IUCN had explained that, in his estimation, Maurice never did anything without a predetermined objective. But what could that be in my case?

The next day, Maurice gave a rousing opening address as part of the ICIMOD inauguration ceremony. The front rows of the auditorium were occupied by the King of Nepal, the prime minister, and a bevy of ministers, foreign diplomats, and ranking bureaucrats from UNDP, USAID, GTZ, SATA, IUCN. I was delighted to hear Maurice use many passages from the twelve pages of comments that had been hastily sent from Bern to Vancouver the previous summer.

In early 1984, Maurice contacted me several times on mountain affairs, usually by telephone to my home in Boulder. Then followed a surprise. He had business in Denver and would like to use the occasion to meet my family. Could he invite himself to dinner? Of course, I was mystified but didn't hesitate. Pauline was much less enthusiastic. How could I expect her to prepare a dinner for a high-powered multi-millionaire of truly global reputation and fame? Couldn't I take him to a restaurant? So I pleaded to her that it appeared to me that he wished to meet the family and, in any case, how could I take a multi-millionaire out to dinner?

Maurice arrived in Boulder at the beginning of March, having booked a room at the newly refurbished Boulderado Hotel downtown. I acquired high quality wine, both white and red, that we could scarcely afford. All was eventually set for a grand family dinner, except that Nadine was in Ottawa, a student at Carleton University, and Tony, similarly, in Rochester, New York.

I picked up Maurice at his hotel early in the evening. On the drive back to our house he apologized for having forgotten to let us know that he was diabetic! I thought of the rich layer cake Pauline had prepared and the expensive wine I had bought. Pauline met us, at her best and most gracious, at the door. While Maurice was introducing himself to Colin and Peter, I managed to whisper to Pauline, "Sorry, but he is diabetic." Her face demonstrated a mix of feelings, but she managed to substitute a fresh fruit desert, without cream, and Maurice promised to take a half-glass of white wine.

Conversation over dinner flowed freely although it took something of the form of an avuncular enquiry into our family welfare: How did we feel about fitting into living in the States? Were there many remaining ties with Canada? One highlight was Maurice's response to Peter's admission that, at 17, he was concerned about which universities he should consider. Maurice shocked him by saying he should not put too much energy into obtaining a university degree – he, Maurice, had left school at age 14, had never attended university, yet had recently received his 27th honorary doctorate! The somewhat awkward silence that followed was relieved by Maurice asking both Pauline and me what such a fine Canadian family was doing living in the States. Wouldn't we rather be back in Canada? When I answered in the affirmative, he said he would see what he could do about it.

Two weeks later, Maurice telephoned from Ottawa. Could I be there the next week as he would like to introduce me to Ivan Head,[89] president of the Canadian International Development Research Centre (IDRC) and some of the senior staff of CIDA. Of course, I went to Ottawa, staying for two nights with our friends Peter and Ellie Johnson at Manotick.

Early morning on the Wednesday, I witnessed Maurice Strong in action in his Ottawa headquarters. Next there was a visit to Hull (now Gatineau) and coffee with some of the upper echelons of CIDA. Then I lunched with Ivan Head and Maurice at Le Café, the restaurant attached to the National Arts Centre. I realized that this was the most critical part of my visit. We finished lunch before any serious talk and then discussed at length mountain development problems facing Third World countries. Mr Head seemed especially interested when I introduced the theory of Himalayan environmental degradation and the accompanying criticism of aid agencies that seemed uninterested in fact-finding before spending large sums to solve vaguely defined problems.

Two days later, when I was back in Boulder, Maurice telephoned again. He and Ivan Head wanted to create a mountain research institute attached to a Canadian university. They would like me to serve as its director. If I were in agreement, would I send a proposal following the guidelines that he then recited – a graduate research and teaching institution for both Canadian and Third World students, emphasizing post-graduate teaching and research, and with an interdisciplinary faculty? There should be a close link with UNU, and could I move with me the IMS headquarters and the journal *Mountain Research and Development*. If I was in agreement in principle, he would immediately contact the president of a Canadian university who was a friend of his. There would be generous funding from CIDA and IDRC that could be justified as a federal expenditure because the objectives would fit in with Canada's overall international aid objectives.[90] Research on Canada's own mountain problems would rely on alternative sources of funding.

For the next several weeks, I was in a state of great anticipation. A whole new opportunity had opened up. At the same time, I was moving ahead with the UNU mountain project, publication of the journal with Pauline, and work on a position paper on the Himalayan issue. I also had a responsibility to continue teaching undergraduate and graduate courses in Boulder as my sole source of financial support.

Suddenly the plan collapsed. On 1 July 1984, Maurice made the fateful telephone call from Ottawa. He said that it was only the second call he had made that morning. The first had been to Mr Trudeau offering his resignation as advisor. In the federal general election of the day before, the Liberals had been heavily defeated and, without support from Mr Trudeau and a Liberal government, there was no chance to create a new mountain institution in Canada. He expressed great disappointment to have built up expectations, but asked me to keep in touch as there would be other occasions when we could work together.

Mohonk II: Refinement of a Political Mountain Agenda

The best way to overcome such a disappointment proved to be an additional burst of energy into the UNU mountain project and the journal *Mountain Research and Development*. This, in turn, led back to Mohonk Mountain House. Intermediate discussions had begun with five important collaborators: Michael Thompson, Larry Hamilton, David Pitt, David Griffin, and Sun Honglie. The fortuitous contact with Michael Thompson has already been explained (Chapter 10). Larry Hamilton had been corresponding with me

for several years following a UNESCO report he had been asked to prepare while still on the faculty at Cornell University. Now as a senior scholar with the East-West Center, University of Hawaii, he was able to ensure the Center's co-sponsorship of Mohonk II. David Pitt was a New Zealand professor of social science then with IUCN working on problems of mountain women and young people. David Griffin, from Cambridge and Canberra, was the director of the Nepal-Australia Forestry Project and ensured a flow of highly relevant manuscripts (Mahat et al. 1986a 1986b 1987a 1987b). Sun Honglie, one of our first UNU post-doctoral fellows, was now vice-president of CAS and vital for the 'China' contact. Bruno was involved in practically all our mountain activities. Bruno, Larry, David Pitt, Honglie, and Michael formed the Mohonk II organizing committee and, with me, chaired most of the conference sessions.

However, a significant step was to be taken before all the foregoing contacts and developments could be combined into effective action. As explained earlier, UNU had accepted my suggestion that the large flow of results from the research in Nepal and Yunnan should be summarized as a substantial position paper entitled "The Himalaya-Ganges Problem" (Ives and Ives 1987). A one-time supplemental grant was made to the International Mountain Society to facilitate this. The growing number of new contacts prompted me to include collation of the large number of submitted manuscripts, some of which were published in *MRD*. Ideally, this accumulation of new research results could greatly enlarge the basis for an international conference. Thus, when Brad Snyder telephoned from Mohonk Mountain House to ask when I was going to respond to the Mohonk Preserve's proposal for a follow-up conference, I answered immediately, "whenever your board of trustees are amenable to a conference with the title 'The Himalaya-Ganges Problem.'" Approval was immediate.

After discussions with Bruno, Larry, and Walther Manshard, I suggested April 1986 and immediately asked Maurice Strong to be honorary chairman. As he was then working in New York as UN under-secretary-general, it was for him a short drive to New Paltz. He agreed to join us. There remained a heavy workload before we all met the following April at that delightful hotel close to the Shawingunks.

International Challenge to Himalayan Mountain Myths

Mohonk II in April 1986 attracted participants from Australia, Bhutan, Nepal, Canada, China, France, Germany, India, New Zealand, Nepal, Norway, Pakistan, Switzerland, the United Kingdom, and the United States. Sir Robert G.A. Jackson, senior advisor to the UN secretary-general, attended as one of our several distinguished guests. The Aga Khan Rural Support Programme, ICIMOD, IDRC and CIDA (Canada), SATA, UNDP, and the World Bank sent senior representatives and provided various forms of support, as did the governments of Bhutan, China, India, and Nepal. The BBC TV and Radio World Service and French TV provided news media coverage. A special success was to persuade Shri Sunderlal Bahuguna, leader of the Chipko Movement, to attend. My first attempt had met with a refusal because he would never accept national government or UN funding which he thought would taint the purity of his peoples' political activity. Nevertheless, he was able to accept travel support from the IMS and Mohonk Preserve covered his local expenses. Finally, with additional UNU support, I prepared 20 large (40 cm by 40 cm) colour prints of Hasselblad photographs from the Himalaya and the Yulong Xueshan, Yunnan, for mounting and display at the conference, mainly to attract the news media.[91]

On my arrival at Mohonk Mountain House, a day early, I was invited to coffee by the hotel manager. She went over the list of participants with me. Noting the visitors from all the Himalayan countries, including government, NGO, and academic representatives, she asked whether they should be searched for weapons. Did I really expect a peaceful debate? Sunderlal had also arrived early, in full flowing robe, and with a large placard declaring that the forests of the Himalaya belonged to the mountain people and not to government and commercial exploiters. His appearance as an Indian mystique (a latter-day Mahatma Gandhi), complete with long beard, had especially impressed her staff. I assured her that he was an active pacifist.

The State University of New York, New Paltz, allowed use of a large auditorium for the opening ceremony and I gave an evening public lecture on "The Himalaya-Ganges Problem" (Ives 1987). This set the stage for four days of formal presentations, discussions, and informal debates interspersed by conducted walks in the Mohonk Preserve. The discussions were very spirited with much contention although the early concern expressed by the hotel management about the possibility of open warfare was easily allayed.

On the Saturday morning, Maurice Strong chaired the closing session. The main objective was to ensure unanimous passage of a series of resolutions. Given the widely conflicting interests and competing political associations of many of the participants, Maurice displayed his diplomatic skill. Despite a number of disagreements, deft word changes and explanations ensured that there were no abstentions nor persistent objections. The resolutions are reprinted as Appendix II to this volume.

I returned to Colorado eager to work with Pauline to ensure early publication of the conference proceedings. I realized that Mohonk II had been a considerable success, but its full implications only became apparent over the course of the next several years. The proceedings were published within little more than a year (Ives and Ives 1987). The book *The Himalayan Dilemma: Reconciling Development and Conservation* (Ives and Messerli 1989) followed, although it was a decade later that the term "Mohonk Process" came into being (Thompson 1995; Forsyth 1996). This was a tribute to Mohonk II that had set the stage for the challenge of several other major international orthodoxies, such as "desertification."

Of significance was the very sparse interest shown by the news media. Although the media representatives who had been invited certainly produced highly supportive reports, including a book on Nepal (Pye-Smith 1990), the worldwide coverage that we had hoped for did not materialize. It seemed that only predictions of catastrophe and death on a large scale would catch widespread attention. The "news interest" did not remotely compete with that achieved by Erik Eckholm.

The Mohonk link with Maurice Strong, however, was possibly the single most important contribution. It became an effective pathway to the Rio de Janeiro 1992 Earth Summit and inclusion of the "mountain chapter" in Agenda 21. Yet it could be argued that, without the much earlier intervention of the ghost of Sir Wilfred Grenfell and a church Sunday School in Grimsby, England, none of this may have materialized.

Fig. 101. This young girl is being encouraged by her mother to become accustomed to carrying a basket, at first empty. As she reaches her teens it will contain half the weight of that of her mother's load. The next step in her education will be that of a 'beast of burden' for the rest of her life. What can be done to change her prospects?

Fig. 102.   Sherpani heavily laden with firewood. Do the tales of environmental degradation in the Khumbu Himal, and in other mountain regions, depend upon this kind of deforestation?

CHAPTER 13

Mountain Road to Rio
The UN Earth Summit 1992

One evening in March 1990, I picked up my telephone to receive a surprise call from Ottawa. It was a surprise because it was from the son of Maurice Strong. He had been asked by his father to let me know that he was setting up campaign headquarters in Geneva as secretary-general of the planned UN Conference on Environment and Development (UNCED, or the Earth Summit), scheduled to assemble in Rio de Janeiro in 1992. Among the tasks ahead for him was to oversee the formulation of a long range environmental plan for reversing the negative trends that were besetting the entire world. Eventually this objective would be framed as Agenda 21. He determined that it should provide a prominent place for the mountain issue. Would I immediately alert my colleagues in the mountain advocacy group to assist?

In the years before this urgent request, our mountain group had become increasingly concerned about how to attract the attention of key players on the world stage. Several lines of field research, workshops, and publications had been highlighting this need, especially since the second conference at Mohonk Mountain House in April 1986, but as explained in Chapter 12, we

had failed to attract the news media.

During a 1986 meeting of the IGU mountain commission in Barcelona and the Pyrenees, Bruno and I became acquainted with Yuri Badenkov, assistant director of the Institute of Geography, USSR Academy of Sciences. He invited us to accompany him to Tajikistan and the Pamir in August 1987 and, in so doing, became a core member of our group. The excursion also led to an expansion of the UNU mountain project into the Pamir, but the region became enmeshed in the Tajikistan civil war of the 1990s and our efforts had to be abandoned.

Also in November 1987, I was invited as one of twenty international experts to take part in a study week organized by the Pontifical Academy of Sciences. The meeting took place in the magnificent setting of the Vatican Gardens and included personal interviews for the participants with His Holiness John Paul II. The objective of the study week was to develop "a modern approach to the protection of the environment."[92]

I was asked to make the case for the mountains. In effect, it was the first occasion when mountains were represented at a high-level international conference on the environment. My presentation resulted in a vigorous debate over the Himalayan deforestation controversy. My chief opponent was Dr Mustafa Tolba, who had succeeded Maurice Strong as director-general of UNEP.[93] But the most memorable moment was my private meeting with His Holiness John Paul II. I was moved by his warm personality, his profound enthusiasm for mountains, and by how extensively he had been briefed on my own mountain commitment. I was equally encouraged by his promise of support.

Our group of academic mountain advocates had also discussed the need for a much stronger base for our growing political agenda and support for long-term mountain research. UNU had recently established a semi-autonomous institute, the World Institute for Development Economics Research (WIDER), based in Helsinki with financial support from the Finnish government. Using WIDER as a prototype, I asked Walther Manshard to support the preparation of a formal proposal recommending a similar institutional arrangement for mountains. This was agreed, and Larry Hamilton was able to have the East-West Center of the University of Hawaii act as host for a small workshop in August 1988. A 120-page document was produced and submitted to UNU. For the first time, Walther was unable to help, presumably because we lacked a cooperative country, such as Finland in the case of WIDER, to provide the long-term financial guarantee. This brought home again the sense of loss over the collapse of Maurice Strong's efforts to create a Canadian mountain institute.

From my personal point of view, the political agenda finally began to take form with the March 1990 telephone call from Ottawa. Maurice Strong's confirmation as secretary-general of UNCED and his move to Geneva provided us with a specific objective – to present a mountain platform at Rio. But how could we achieve such an ambitious plan?

From this beginning, we were joined by a combination of Peter Stone, editor of the UN broadsheet on environmental affairs, and Ruedi Högger, former vice-director of SDC, who had been working closely with Bruno and me since our early UNU activities in Nepal. They had made a joint personal approach to Maurice Strong in early 1990, shortly after he had established his headquarters in Geneva.

With UNU and SDC funding, a strategy meeting was held in Canton Bern the following December. It took place in Appenberg, in the Emmental home-from-home of my family during the sabbatical year of 1976–77 and now enlarged as a small resort and conference centre. Ruedi chaired the meeting that produced a first outline of a strategy. Bruno, Larry, Peter,[94] Yuri, Frank Tacke, the current director-general of ICIMOD, and I were the core members, aided part of the time by Hans Hurni and Hans Kienholz.

We needed an official-sounding title for our group and after deliberation chose the name *Mountain Agenda*.[95] The strategy included plans for a substantial book, which became *State of the World's Mountains: A Global Report*, and a brochure "*Appeal for the Mountains*," to be presented to busy diplomats and journalists who would probably never have time to read a book of this length. Preliminary tables of contents were outlined and potential writers identified for the various chapters. Requests were sent to the Swiss and German governments for financial support. Peter Stone would guide the book through final editing and layout and find a publisher. I had the task of selecting possible authors, in addition to those at Appenberg, and helping with the writing and editing.

To produce a major book within 18 months with 300 copies delivered to Rio in time for the Earth Summit was an ambitious task. It was a great relief that our appeals to the Swiss and German governments produced grants totalling US$300,000.[96] Following the Appenberg strategy meeting, we launched an extensive letter-writing campaign identifying our intentions and appealing for support. A follow-up meeting was scheduled for April 1991, by which time we hoped that a substantial part of the book, at least in first draft, would be available. The April meeting took place in the small mountain village of Aeschi above Lake Thun in the Bernese Oberland, where Jayanta Bandyopadhaya and Martin Price were added to the core members. By the end of the meeting, we felt that we would meet the publication deadline, but only just.

By the beginning of 1991, many organizations and groups with the intention to carry the mountain message to Rio were being identified. For me, the next significant step was taken in December 1991: a telephone call from Walther Manshard asking me to go to Tokyo to make a presentation to the UNU Council during its annual general meeting and to request formal support for the mountain mission. I was already committed to another vital meeting in Bern, but Walther said he knew this and would purchase a round-the-world air ticket that would enable me to make the essential UNU Council meeting,

then continue to Zurich, on to a meeting with Maurice Strong in Geneva, and back to Bern on one of those reliable Swiss trains – a timetable that allowed for no stopovers! I think it would have required less energy to reach the South Col.

A snowstorm had closed Zurich airport just after my flight from Tokyo had landed. I adjusted to take the train to Geneva that allowed a three-minute window for Yuri Badenkov to jump aboard in Bern. Maurice had a taxi waiting for us at Geneva railway station. I was able to greet him with the news that the UNU Council had asked me to represent the university in Rio with a colleague of my choice. As Bruno was to be a member of the Swiss delegation, I asked Jayanta Bandyopadhyay from India to assist. Maurice promised to sign any support letter that I prepared and to help in any way possible.

There was still a long journey ahead. The UN had required the setup of four preparation committees (in UN-speak, "prepcoms") that would progressively formulate the overall plans for Rio. The first two were strictly *in camera*. The third and fourth were more open to outside comments and questions: one in Geneva and the final one at UN headquarters in New York.

Precom 3, held in Geneva in October 1991, had been a success for our cause because support had been received for a "mountain chapter" to be included in Agenda 21, which was now approved in principle. Nevertheless, a large group of small island UN member countries had formed a voting bloc, and the United States had made it known that it would oppose a second bloc of small countries, such as the mountain countries, that together with the islands might work to thwart U.S. interests. In addition, one of the FAO delegates had appeared to be totally opposed to a specific mountain chapter in Agenda 21. His argument, that we had heard before from other sources, was that issues such as soil erosion, forestry problems, and other topics that would enter any possible mountain chapter, would be covered in other chapters and would thus be redundant. To obtain permission to speak, Bruno and Ruedi had to be made temporary members of the Swiss delegation in order to counter this argument. The struggle was eventually won, and the first draft of the mountain chapter was written. Although the chapter was subsequently modified beyond recognition, the chapter title was retained intact.

The fourth and final prepcom in New York could still see the mountain chapter rejected although the provisional text was tentatively entrained into the UNCED vehicle and was brought for presentation to the prepcom. This was to be the final test.

Switzerland was not actually a member of the UN until 2002 although prior to that it had held observer status with substantial offices in New York. The Swiss delegation, with Jean-François Giovannini as its head, organized a reception for representatives of all the mountain country UN delegations – from Austria to Nepal, from Norway to Ecuador, Iceland, and Japan. Jean-François phoned me asking that I and Jayanta attend. He explained that our task was to help develop a common front among the mountain countries in preparation for the debate scheduled in the UN General Assembly the next day.

We addressed the mountain country delegates, urging them to accept the draft chapter as it was, despite the obvious improvements that could be justified. Only the title was important, if we were to get it past the debate at the UN the following morning. Thus, the objective was to limit and control as much as possible the discussion in the UN. We were unsure of the stand the FAO might take, but at least they only would have observer status along with the other UN agencies and could not directly address the General Assembly. We also had to anticipate U.S. opposition on purely political grounds, and several other member states were still arguing that a mountain chapter would be unnecessarily repetitious and therefore not needed. Jean-François reasoned that the actual chapter wording could be changed, but only *after* Rio when the chapter itself, by its title, would be tightly bound into Agenda 21.[97]

Jean-François also coached us to explain that the various mountain country delegations, who would be scattered in alphabetical order throughout the General Assembly, must try to remain in hand-signal contact. This would facilitate use of the standard parliamentary blocking system in the event that potentially hostile questions were raised. The primary objective was to achieve agreement in principle for the general concept of a mountain chapter; details could be left until later.

The final nicety of the evening was that the head of the Ethiopian delegation, our mountain colleague Dr Tewolde,[98] asked if I would join him on the floor of the General Assembly as an ad hoc member of his delegation.[99] This would enable me to prompt him, if and when necessary, and help him send signals to other supportive delegations.

I must admit that I was rather pleased with myself, red-headed and blue-eyed, to be a member of such a distinctive delegation. It was a memorable experience and, despite a few anxious moments, the discussion was restrained the next morning. Chapter 13 was passed in principle, ready to be transferred to Rio.

Success in Rio

The Earth Summit was staged with worldwide fanfare in Rio de Janeiro, 3–14 June 1992. As the majority of the world's heads of state were expected to attend, the Brazilian security arrangements were probably the most intensive ever brought into play up to 1992. The main route from the principal hotel area overlooking the spectacular Copacabana beach, was lined with the military, with tanks stationed at all the main crossroads. The vast new conference centre had a double, heavily guarded defence perimeter. Jayanta and I were fortunate to be invited to ride with the Swiss delegation.

I have a vivid recollection of entering the crowded main hall to find Maurice Strong talking with a stranger (at least to me) in the only patch of open space. I pushed through the crowd to congratulate him on his remarkable triumph. He introduced me to Mr Lee Kuan Yew, prime minister of Singapore, explaining to him the outstanding support he had received from his mountain colleagues. Strong added that he was confident the mountain chapter would pass by acclamation on its scheduled presentation in three days' time. I voiced the view that without any controversy we would obtain scant news media coverage. He responded, "Jack, don't you think I will have enough controversy on my hands without any more from the mountains."

The headquarters of most delegations were scattered throughout the conference complex. Bruno, Jayanta, and I found our way to the Swiss offices, where a large number of our books, the *State of the World's Mountains* and the *Appeal*, awaited us. Our immediate task was to distribute copies, together with a covering letter, to as many of the delegations as possible in the time available. We decided to concentrate on 75 member states whom we assumed would have a strong interest in mountains, dividing the task into three piles of the volumes. My group included Iceland, the Vatican, the Scandinavian countries, UK, Canada, USA, Japan, and China, amongst others. It was a big task, and Bruno was heavily preoccupied as a member of the Swiss delegation; Jayanta and I were essentially free agents, although we barely finished and still had to reach the "non-mountain" majority.

We used a lunch break to enter the usually closely guarded General Assembly room when we surmised that it would not only be empty, but that the security personnel also would be at lunch. Dividing the seating arrangements into equal halves and with two loaded trolleys borrowed from the Swiss delegation's offices, we set about placing copies of the *Appeal* and the covering letter on every desktop. Fortunately, we started from the centre and worked outwards. I had almost completed my half when I was accosted by whom I assumed was plainclothed police and firmly escorted to the security office on the edge of the assembly hall, where I was interrogated.

Much to my surprise, one of my interrogators identified himself as CIA. However, my explanations were accepted, and I was let off with a reprimand and told to leave the hall. I couldn't help laughing to myself as I realized that, while this was going on, Jayanta was completing the task unhindered. No effort was made to remove the documents, so our ad-campaign worked, although whether it had been necessary we could never know.

As Maurice had predicted, the inclusion of mountains as Chapter 13 in Agenda 21 was approved by acclamation. And as I had predicted, the lack of controversy meant little news media coverage. Two days later, Jayanta and I, accompanied by UNU Rector Professor Heitor Gurgulino de Souza, met with a group of journalists. While there were many questions, it all seemed a formality. We did manage some short notices, although the world news media had bigger fish to fry than our mountains – the same problem that had faced us from the beginning. Nevertheless, the Mountain Agenda had triumphed at Rio.

Rio-Plus-Five: UN Headquarters Review, 1997

Despite the euphoria generated by our experience in Rio, we quickly came to realize that there was much more to be done. FAO had been appointed as lead agency for putting the mountain chapter into effect, and leaders of our mountain advocacy group were soon called to Rome for yet more planning and implementation. Here we met Jane Pratt of the newly invigorated "Mountain Institute" (based in West Virginia, USA), who was to play a vital role later on.

The Rome-FAO meeting led to a conference in Lima in 1995 hosted by the International Potato Institute when The Mountain Forum was founded. Jane Pratt was the principal organizer, and the forum constituted a worldwide mix of individuals, NGOs, institutes, and bilateral and UN organizations with the general purpose of spreading the word and publicizing the urgent need for new policies for sustainable mountain development. It proved an important attraction for representatives of many groups of indigenous mountain peoples.

The UNCED coordinating council had set 1997 for review of the progress made with regard to six of the Agenda 21 chapters, and this review included Chapter 13. This shaped up into a requirement, basically for our own group of mountain academics, to produce a substantial report that turned into yet another book.[100] We decided that, in producing our report for the UN in the form of a book, we would also reach universities and other institutions worldwide. This became another desperate writing and editing task with an almost impossible deadline (June 1997). Once more, the Swiss government contributed substantial funding with a modest assist from IDRC, facilitating editions in English, French, Spanish, Italian, and Russian. Production required more than 30 principal authors and more than 70 subsidiary writers. Only the English edition could be printed in time to take to UN headquarters because of the lengthy translation process.

The book launch was held at UN headquarters in June 1997. It also provided a platform for the UN delegate of the Kyrgyz Republic, Mrs Zamira Eshmambetova, to make a request for the UN to declare an international year for mountains. Her appeal was reinforced the same evening over dinner with Bruno and Jean-François.

The rush to complete the book and prepare a speech for the launch in New York coincided with my retirement from the University of California and transfer with Pauline back home to Ottawa after 30 years in the United States. This retirement also saw the completion of two decades of editing and production of *Mountain Research and Development* and its transfer to Bern. Hans Hurni took over the task of editor-in-chief and received greatly increased funding from the Swiss government.

Declaration of 2002 as the International Year of Mountains

By 1997, the ripples from the Rio Earth Summit had spread across the entire world: government agencies were being established, and from our particular point of view, NGOs purporting mountain expertise were multiplying while funding for applied mountain research was augmented. Most important, perhaps, was that indigenous mountain groups were becoming involved. One of the off-beat responses to the overall success of UNCED was the widespread use of the terms "eco-friendly", "green", and "environmental sustainability"; they even became household phrases and, insidiously, were exploited by the commercial world.

Our 1997 book was well received and the Kyrgyz proposal, strongly backed by the Swiss government, acquired widespread support in the UN General Assembly, culminating in

approval in November 1998. The date of 11 December 2001, was chosen for the formal declaration of 2002 as the International Year of Mountains. Thus a large group of mountain advocates met just two months after the infamous 9/11 attack on New York City and rather close to the desolation of Ground Zero.

Adolf Ogi, a former president of Switzerland, opened the meeting with the resounding notes of an alphorn played by a traditionally dressed Swiss alpinist. A series of keynote speeches followed: Murari Raj Sharma, acting president of the UN General Assembly; Jacques Diouf, director-general of FAO; and Kurmanbek Bakiev, prime minister of the Kyrgyz Republic, were the opening speakers. I addressed the audience on behalf of Hans van Ginkle, UNU rector, who was unable to attend in person. The ceremony concluded with a lively round-table discussion that emphasized the integral importance of the mountain lands to the well-being of the overall world environment and the livelihood of all people.

This was certainly an occasion for celebration. Bruno, Larry, Jane Pratt, and I, and our spouses sensed the fulfilment of decades of concerted effort. But what had we achieved? By "we," I refer the dozens of individuals and organizations without whose collaboration the cause of the mountains would never have been heard in Rio in 1992.

Now the cause has been taken up again during the Rio+20 Conference, also held in Rio de Janeiro, in June 2012. In this huge enterprise, that engaged most of the UN member countries and many NGOs, an enigma arose – the struggle between the short-term political-economic crisis and the longer-term environmental imperatives (Chapter 17).

Fig. 103.   The highest summit of the Jade Dragon (Yulong Xueshan) from what in 1985 was close to the centre of the old town (Dayan) of Lijiang. The tourist juggernaut was still a few years away.

CHAPTER 14

Return to China 1991–1996
Mountain People of Yunnan and Mass Tourism

In 1991, I received a request from the Ford Foundation to organize an international reconnaissance of northern Yunnan in association with Professor He Yaohua, president of the Yunnan Academy of Social Sciences. It was proposed that, ideally, the group should include scholars with mountain experience from several of the developing countries, and especially women. The objective was to report on the feasibility for a multi-year study of the anticipated impacts of accelerating modernization on the local environment, with emphasis on ethnic minorities, particularly women. A staff member of the Beijing office of the foundation, Dr Nicholas Menzies, would accompany us as well as Professor He and several of his staff.

Selection of a suitable team was greatly facilitated by the extensive mountain contacts that had been developed over the years; we soon had colleagues keen to join us from Peru, Nepal, Tajikistan, Canada, and the United States. The University of California, Davis contingent included Janet Momsen, a well-known authority on the position of women in developing countries, Margaret (Peggy) Swain, an anthropologist with experience

in southern Yunnan, and myself. Muazama Burkhanova, whom I had met in Dushanbe, was also able to join us. A particularly important recruit was Wu Ga, a doctoral degree student at Michigan State University. She was Yi and would be able to provide Yi, Mandarin, and English interpretation for our anticipated interviews with Yi village women so that we would not be exclusively dependent on our Chinese hosts.

From all points of the compass, we converged on Kunming in September 1992 to meet Professor He and our other Han Chinese and Naxi hosts. The following road trip covered a large area of southern Sichuan and northern Yunnan, and it included visits to many villages and several small towns. Along the way, I felt we were being too closely monitored by a young Han lady member of our party who was fluent in Yi and Naxi and had excellent English. In one of the villages, we managed to distract her attention and so enable Janet, Peggy, and Wu Ga to obtain unsupervised interviews with Yi women. This prompted me the following February in Beijing to test our concern that we might not be allowed unrestricted interviews with minority

J. D. Ives, *Sustainable Mountain Development*, https://doi.org/10.1007/978-3-030-96029-2_14

Fig. 104. The University of California-Yunnan Academy of Social Sciences team (1993-1995) standing in front of a Naxi temple.

people if and when we began a serious long-term investigation.

The Lijiang-Yulong Xueshan region was chosen for the new research, in fact it became a follow-up to the 1985 CAS-UNU fieldwork. UNU promised support as co-sponsor of the project with the Yunnan Academy of Social Sciences. The Ford Foundation asked me to attend a conference on Chinese minority peoples in Beijing in February 1993 to present an overview of our findings.

I concluded that presentation in Beijing with a candid account of our surreptitious and unsupervised Yi minority interviews. As the Chinese central government minister of minority affairs was present, this was, while perhaps risky, a sure test. I recounted our stratagem for obtaining candid reactions from Yi women to the new Chinese policy of one child per family (two for minority families). One Yi lady had explained that, as she already had a daughter, she was faced with a dilemma: the risk of another daughter, that might prompt her husband to abandon her, would condemn her to a secret annual abortion unless she could satisfy his wish for a son.

Following my talk, I was escorted to a private interview with the minister. After the invariable tea formalities, a foreboding silence ensued. Finally I spoke. I first apologised if I had inadvertently caused any offence, explaining that, as an apolitical academic, I had not intended to open up any delicate political issues but had felt obligated to describe our findings. The minister explained that I was under a misapprehension. The government recognized their own need for such unbiased information and that they would never be able

Map 8.   The Lijiang region of northwest Yunnan, China. The Naxi and Yi villages that were part of the field investigation. The proposed trekking route would pass through the Tiger Leap Gorge and connect with the small villages.

to obtain it themselves, while we, as outsiders, could. He insisted that he would strongly support our project.

The minister's response was most reassuring. There would have been no justification for encouraging graduate students to invest time on thesis topics only to find they were stopped midway. In practice, it seemed that we would have unlimited opportunity to undertake serious anthropological research that many Westerners would have regarded as politically highly sensitive, if not completely proscribed. The Ford Foundation approved a three-year research proposal and we returned to Kunming the following September.

Lijiang and the Yulong Xueshan, 1993, 1994, 1995

Three post-monsoon visits to the Lijiang-Yulong Xueshan region led to research on a scale adequate for significant results. The team included Janet, Peggy, and myself, together with Seth Sicroff and Lindsey Swope, UC Davis graduate students who were to undertake fieldwork for their theses. Professor He accompanied us for most of the 1995 season and provided extensive support throughout. Our academy colleagues were Yang Fuquan (Naxi), Feng Zhao, Zuo Ting, Du Juan, and Mu Liqin. There were jeeps and drivers from the academy and assistants from the Chinese National Tourism Office in Lijiang. Xie Ye, governor of Lijiang County, and Mu Rong Xiang, governor of the prefecture, became ex-officio members of our team. As a reciprocal gesture, Xie Ye bestowed on me the title of "Special Advisor to the Governor for Tourism and Sustainable Development" (informally, of course). This led to a close relationship that eased our way in contacts with village elders and later gave us some influence on development policy for both the city and the region.

Janet and Peggy had produced an outline questionnaire before leaving California for China. It was set up to assist in determination of rural village household workload, subsistence needs, annual work calendar, reproductive history, and male–female division of labour. The draft was discussed by the entire group at the academy headquarters, modified, and then field-tested over three days in the small village of Danouhei, in Shilin (Stone Forest) County, located about 50 km southwest of Kunming. Modifications were introduced, and several hundred copies of the revised questionnaire were made in Kunming in time to accompany our two-day drive to Lijiang.

In Lijiang, we were housed in the best available hotel, which we used as our base of operations throughout the three long post-monsoon visits. The first task was to select the villages for detailed investigation and, at the same time, provide the entire crew with a good overview of a large section of the region. We made a three-day trek through the Tiger Leap Gorge with two nights in "Walnut Grove" at a quaint improvised tourist lodge. The proprietor, who had adopted the name Sean, became a helpful friend.[101] He had been badly burned as a small child, losing one hand and severely damaging the other. He had the good fortune to learn English as part of a government programme available to children who were physically impaired. Thus, long before most others imagined that English-speaking tourists would flock to Lijiang, he was one of the very few locals to acquire this critical language skill. Sean also had the presence of mind to recognize that the Tiger Leap Gorge would become a major tourist trekking destination; already by 1993 he had constructed a sparse yet acceptable lodge at a most attractive locality in the gorge. He had a Han wife and three mischievous little girls, two of whom were old enough to assist and who became tourist attractions in their own right – they would wait on tables and sing "Frère Jacques" in five languages.

We selected four villages for intensive survey and one for a brief investigation and made cursory visits to several others. It proved very useful to have Governor Xie Ye's official approval when, during our introductory enquiries, we approached village elders. Regardless, we were greeted enthusiastically wherever we went.

Of the many Naxi villages, we chose Wenhai, where I was welcomed as an old friend.[102] No formal permission was needed here. The other two Naxi villages were Yuhu, the place Joseph Rock had made his headquarters in the 1920s and 1930s, and Jiazi. Our main Yi choice was Hei Shui – in effect, three tiny hamlets situated on the edge of the Yulong Forest Preserve. We also selected the small Yi village of Ru Nan Guo, an especially poor settlement about an hour's walk northwest of Wenhai. The study site locations in relation to Lijiang and the Tiger Leap Gorge are shown in Map 8:

- *Wenhai: Naxi;* alt. 3110 m, a day's walk from the nearest road head.
- *Yuhu: Naxi;* alt. 2710 m, 3 km by tortuous track from the road system, but driveable under most conditions.
- *Jiazi: Naxi;* alt. 2150 m, accessible by 10 km 4-wheel drive track, but not negotiable for much of the rainy season.

- *Hei Shui: Yi;* alt. 2830 m, close to main road leading north from Lijiang and within easy walking distance of the Forest Reserve headquarters, itself at the head of the trail leading to the "love-suicide" meadow.
- *Ru Nan Guo: Yi;* alt. about 3000 m, giving distant access to the southern entrance to the Tiger Leap Gorge. Here we interviewed only the purported poorest and wealthiest family.

This range of Naxi and Yi settlements gave us an opportunity to test for any correlations between economic prosperity and altitude (length of growing season) and for distance (difficulty of access) from Lijiang. Use of the extensive questionnaire, which required multiple interpretations, was time consuming: Yi to Mandarin to English and Naxi to Mandarin to English and the reverse, and occasionally Yi to Naxi to Mandarin to English. Interviews were conducted by two or three of our party who were usually invited into the interviewee's house. Most were women. Invariably there was a cordial, even festive, atmosphere. I have vivid recollections of sitting cross-legged on mud floors facing the embers of the central fireplace eating roast potatoes and drinking tea.

The language barriers were not entirely overcome, but we obtained a large amount of data. Responses of the women to the more intrusive questions about fertility, reproduction, even birth control, were surprising and would have been unthinkable in Sherpa and Tibetan households. One outstanding occasion needs to be recounted.

In one of the Yi villages, the question about the approach to birth control produced spontaneous laughter. Professor He, initially, had been very reluctant to consider any questions relating to family reproduction. He was eventually persuaded with a promise that we would first very gently "test the water." On this occasion, the Yi lady, after her first burst of laughter at the mention of birth control, leapt up and ran out of the house, leaving us, in a state of suppressed expectation, eating her roast potatoes. She was back within minutes, accompanied by her sister. Amid much giggling, they went through the entire village telling us who didn't, and who did, and how.

We accumulated over 200 completed questionnaires. It was a learning experience for all of us. It demonstrated once more, this time in anthropological terms rather than environmental, how the UN and bilateral agency reports were frequently generalizations and should not be applied to specific cases.

To cite a few of our findings: the Naxi division of labour was very different from what we had expected. There was a remarkable gender balance. The menfolk cooked and looked after children when needed; they frequently carried water and wood fuel. In many Naxi households, the wife had equal, or even dominant, control of how the small cash earnings were used.

In Yi households, there was a lesser degree of equality and more pronounced male domination in major household decisions. Nevertheless, males assisted in carrying water and collected all the wood fuel. This role, however, followed the male tradition of control of the forest domain as hunters. Women were not permitted to touch an axe and were responsible for a much greater share of the largely subsistence agriculture.

Lindsey, assisted by other members of the group as interpreters, assessed the factors influencing rates of deforestation in Lijiang County (Swope 1995). Her findings led to a reassessment of the reconnaissance conclusions drawn from the 1985 visit: the degree of forest conservation varied significantly from village to village, as did illegal logging. However, the 1985 indications of very-long-term forest removal were further substantiated. The manner and causes of current deforestation were extremely complex. Local village leadership and village and family response to the uncertainty and constant changes in official forest policy, ownership, and effectiveness of policing illegal logging were major factors.

Fig. 105.   This has all the appearances of a Naxi matrons' picnic.

An example of the results of the conflicting pressures was derived from Lindsey's studies in Wenhai. This administrative unit comprises two associated villages, each with its independent leadership. The northern village, largely because of dedicated commitment to forest protection, maintained good stands of timber well into the 1990s. The southern village had looked for short-term gain as precaution against the possibility of unfavourable future legislation. The very poor condition of its surrounding forest clearly demonstrated this local policy difference between the two units (Swope 1995; Swope et al. 1997).

There is no doubt that serious attempts were being made by all levels of government to ensure extensive reforestation, which included a direct subsidy to encourage individual families to plant trees on an annual basis. However, the Forest Bureau staff were not adequate to enforce regulations, illegal logging was widespread, bureau staff were too close socially to timber poachers to ensure

Fig. 106.   The traditional Naxi farm house is a massive consumer of timber. Construction was tightly restricted until Mr Deng Xiaoping gained control of the central government of China and introduced new regulations that were much more favourable to the minority peoples.

Fig. 107.   The treeless slope in the foreground is almost identical to that observed by Joseph Rock sixty years before. The Tiger Leap Gorge is lost deep in the middle ground. The unnamed peaks are part of the Yulong Xueshan. This is our approach to the upper trail through the gorge in 1993.

Fig. 108. Tiger Leap Gorge. Laden horses approach the crux of the lower trail overlooking the precipitous plunge to the tiger's rock. Today this treasure of nature and human perseverance has been blasted away to make room for tourist buses.

adequate enforcement, and corruption was most certainly not insignificant.

Despite this situation, we found an overall pattern of long-term change in forest cover over centuries, as opposed to the sudden change trumpeted by ardent adherents of the assumption that extensive deforestation, even to the point of desertification, had occurred since 1950. Far south of Lijiang, the condition of large areas of the Yunnan Plateau, as seen from our road travels and attested by village elders, was disastrous. Deforestation for centuries had led to such serious soil losses that extensive areas of bedrock were exposed (Figures 70–72); this environmental deterioration was certainly not recent.

An indication of the potential damage that could be caused by misinformation was uncovered during our study of the small Yi villages of Hei Shui situated on the eastern edge of the vigorously protected Yulong Forest Preserve. The Yi were definitely cutting trees in 1985, as were other people, including members of the Forest Bureau. At that time a constant line of trucks, almost bumper to bumper, carried logs southward 24 hours a

Fig. 109 [top].　The Yi villages of Hei Shui situated at the base of the eastern slope of the Yulong Xueshan were selected for intensive study. It proved to be the nearest village group to the Love-Suicide Meadow and its inhabitants were the first to realize the prospects for tourist exploitation (see text pages 219, 221).

Fig. 110 [bottom].　Lower Wenhai village looking across the lake in the post-monsoon season so that it is still close to its maximum level (1993).

215

day. Although that had been curtailed by 1992 and checkpoints had been established, illegal logging still occurred. It was expedient for the Naxi and Han Chinese to accuse the Yi. When Governor Xie Ye informed me that he would evacuate the Yi villages, we pointed out that our surveys showed that the Yi had restrained themselves to using fallen dead wood during the previous several years. This observation of ours effectively ensured that they were allowed to remain in their homes.

During our 1993 stay in the Lijiang region, we recorded the rapid development of mass tourism. Hotels were being constructed in the "new" town of Lijiang, and the number of trekkers from Western countries was increasing. By far the majority of visitors, however, were Han Chinese. This provoked discussions about what we had seen happen as tourism developed rapidly in other countries, especially in Nepal. We speculated that it would not be long before the Tiger Leap Gorge became a world-class destination, and

with it the rest of Lijiang County. A sense of obligation arose such that we were tempted to intervene in the hope that some benefit accrued to the remote and impoverished mountain communities. Seth proposed the promotion of trekking tourism, somewhat along the lines that had been developed so effectively by the Sherpas of the Khumbu. His ambitious plan – circumambulation of the Jade Dragon – was quickly endorsed by the entire team; it became an important theme of his thesis that proved a major documentation and assessment of tourism in northwestern Yunnan (Sicroff 1998).

The plan proposed a seven-to-ten-day trek beginning in Yuhu and consisting of a day's gentle walk from Yuhu to Wenhai; two nights in Wenhai; then northward via Ru Nan Guo and down to the southern entrance to the gorge; two days walk through the gorge with an overnight at Sean's guest house; then onward, out of the gorge, across the river to Daju for one night; finally, two days walk along the eastern flank of the Yulong Xueshan via Hei Shui, and so back to Yuhu. This plan would incorporate most of our study villages, thereby ensuring a supportive reception. We also thought it would be a special way to show our gratitude for all the hospitality and assistance we had received.

We discussed this plan with the village

Fig. 111.   Although Wenhai is a Naxi village, our first visit in 1993 resulted in an invitation for the young women and girls from the nearby Yi village of Ru Nan Guo to perform their traditional dances prior to a welcome feast.

Fig. 112.   An irresistible portrait of one of the young visiting Yi dancers.

elders of Yuhu and Wenhai and obtained enthusiastic responses. Next, we acquired two
farms, one in each village, to be equipped as
trekking lodges and helped to set up village
committees, with strong female representation, to operate the lodges and adjunct activities. There were countless possible additions
such as side trips to prolong the visitors' stay:
an ascent of one of the high northern summits of the Yulong Xueshan that was non-
technical, as reported from the climb of Joseph Rock in the 1930s; partial damming of
the disappearing lake at Wenhai to facilitate
production of edible fish, both for meals in
the trekking lodges and for sale on the Lijiang
market; guided walks to observe the colourful
plumed pheasants and, in season, the great
variety of rhododendrons; and so on. Seth Sicroff, the initiator of the proposal continued
to add schemes to the original idea (Sicroff
1998).

Acquisition of the two farms, I thought,
was achieved in a rather ingenious manner. We could not use research grant funds

Fig. 113 [top].   These farm buildings on the
edge of Wenhai were acquired as one of the
centre points for the proposed trekking route
around the Jade Dragon-Yulong Xueshan (see
text pages 216–219).

Fig. 114 [bottom].   An outlying Wenhai farm
at the foot of a luxuriantly forested mountain
slope leading upward to the highest summit of
the Yulong Xueshan. It occurred to us that, if the
proposed trekking route could be realized (see
text pages 216–219), Wenhai could serve as an
excellent base for mountaineering.

to purchase them. However, our budget did permit paying modest wages to local people to convert them into trekking lodges. Villagers' understanding that profit from trekkers' payments for room and board would provide a long-term revenue stream prompted agreement that the unexpected income they received from us for the renovation work be used to purchase the two farms and retain them under village-wide control.

All this was only a beginning. Trails had to be marked, especially along the eastern mountain flank; villagers required training in management; a corps of trekking guides and accident-and-rescue contingency plans would be needed. As an example, we "commissioned" one of Sean's daughters to guide us to a local waterfall, paid her a fee, and declared her a mountain guide. We believed that the planned two-lane highway to allow air-conditioned tourist bus traffic through the Tiger Leap Gorge must be challenged. Finally, we proposed establishment of a Joseph Rock museum in Yuhu (Rock was a sufficiently colourful figure with National Geographic linkage to support such a plan). Governor Xie Ye was most helpful. We were convinced that our participation in Lijiang County affairs would spin off beneficial results for the well-being of the local people beyond its original research objectives. Our overall hope was that success of "trekking the Jade Dragon" would serve as an example for similar developments in the broader region. We were very hopeful of such outcomes, but unfortunately, this chapter must conclude with a tale of disappointment and lost horizons.

Mystery of the Love-Suicide Meadow

I had visited the idyllic sub-alpine meadow *Yunshanping* in 1982 without realizing that we were encroaching on sacred Naxi ground (Chapter 9). By the time our 1985 group camped there, we had learned part of the tale. In 1993, Yang Fuquan revealed the entire story to us. As this is a tale of ancient tragedy leading to a modern day tragedy of an entirely different nature, it is related here in some detail.

Lijiang (or Dayan) has been the centre of the Naxi people for centuries, if not millennia. The Naxi still retain many elements of their traditional customs and matriarchal social system. Until A.D. 1723, control by the Han central government was minimal and the Naxi enjoyed virtual autonomy. After 1723, the new Qing emperor, Yongzheng, ordered conversion to the Confucian principles of the central government. This caused serious cultural conflict between the Han system and the diverse ways of many of the minorities in the southwest of China. In particular, Naxi custom accommodated free choice of marital and pre-marital partners; teenage love relations were open and accepted. Under Emperor Yongzheng in 1723, this pattern was denounced as primitive and barbarian. The Han-style marriage by contract in infancy was enforced.

Despite political suppression, however, remnants of the Naxi traditions survived. In particular, free love before marriage persisted surreptitiously alongside Han-enforced marriage arrangements. The two mutually exclusive approaches to marriage often came

Fig. 115.   One of the many possible day-trips lateral to the proposed main route around the Yulong Xueshan and through the Tiger Leap Gorge.

into conflict as young people were forced into union with partners whom they may never have met. This led to love-suicide pacts between young lovers who refused to be separated.

In many instances, love-suicide plans were made by groups of lovers. In anticipation of their final act, they would flee their families and spend several days visiting beautiful natural locations, camping and feasting in high meadows in the mountains. Eventually they hanged themselves, or jumped from high cliffs. The *Yunshanping* (meadow in the spruce forest, or the Love-Suicide Meadow) became the most famous of several choice locations for this ultimate "escape" ritual.

The lovers believed that after death they would be transported across the high snow-covered mountains to a mystical paradise where they would live forever in perpetual youth and good health, far from the agonies of their former lives. This is perhaps the Naxi version of the myth of *Shambhala*, or Shangri-La, as depicted in James Hilton's 1933 novel *Lost Horizon*.

The suicides persisted, certainly well into the 1930s when groups of up to ten couples are said to have completed the lethal ritual; there are reports of its occurrence until the early 1950s. Naxi people continued to revere the meadow at least up to the early 1990s; it was taboo to whistle in the meadow, for instance, for fear that the spirits of the young deceased couples would lure transgressors into suicidal ecstasy. I would speculate that the relatively small Naxi population (about 300 000), despite a long and resilient cultural history, has been significantly limited partly due to the loss of so many young lives between 1723 and the mid-1900s.

One version of the folktale features a "Queen of the Kingdom of Love-Suicide" who called upon the girl *Kaimeijjiumiji* to join her in paradise, entrancing her with a softly murmured poem. Figure 116 is a photocopy of one of the rare Naxi pictographs that gives an account of the Love-Suicide Meadow tale.

The modern growth of tourism began slowly in the late 1980s, accelerating in the early 1990s. The legend of the beautiful sub-alpine meadow with its spectacular backdrop of the high snow peaks of the Jade Dragon was energetically promoted by the Lijiang tourist bureau. Ironically, the Yi villagers of Hei Shui, whom we interviewed so intensively, were the first to take note of the financial opportunity. Soon, younger Yi women were dressing in their best colourful costumes and offering guided rides on their garlanded horses to spare tourists the steep two-hour walk to the meadow.

After 1990, the flow of visitors, principally Han, expanded rapidly. Yi profits multiplied, more horses were acquired, and more young women began to line up each morning to attract trade. One suspects that the exotically dressed Yi women were as much an attraction for the young Han male tourists as the meadow and its history. With the increase in guides and horses, competition between the Yi women entrepreneurs developed, leading to shouting matches, scuffling, and cut-throat underbidding.

As tourist interest expanded, the 40 km gravel road north from Lijiang was surfaced and a large car park, small hotels, and restaurants were built. In 1994, a chairlift was erected from the parking area to the meadow, which severely curtailed the business of the Yi women. The horses became idle because there was no place for them in the traditional Yi agricultural system, and they were sold at a loss. In 1993, there were more than 10 000 tourist visits; the meadow was trampled and strewn with litter. The tourist bureau efforts to halt the damage were largely in vain.

The meadow itself and the growing number of small hotels and restaurants at the base of the chairlift attracted costumed dancing teams of young Yi women and girls. By October 1995, the investment on the chairlift, financed jointly by a Hong Kong business group and the Lijiang tourist bureau, had been amortized. During the 1996 spring festival, several thousand tourists visited the meadow each day. More recently, the mass tourist attractions of the Lijiang–Yulong Xueshan region as a whole has led to construction a few kilometres farther south of a much longer chairlift and ski resort, a golf club, while the town of Lijiang has been transformed with new five-star hotels and restaurants.

Lijiang as a World-Class Tourist Destination

My final visit to Yunnan was in 1995. In seemingly idle air flight conversation while crossing the Pacific, I talked to my seat companion about my enthusiasm for the Naxi and Yi people and what we were trying to do to improve their situation. As we parted company in Hong Kong, he gave me his business card: he was the senior World Bank administrator for East Asia. This contact soon proved providential.

The 1995 field season was our longest and most intensive. Plans for the circumambulation of the Jade Dragon were forging ahead. In Lijiang, we met by chance a group from Simon Fraser University, British Columbia. They were accompanied by Canadian First Nations Haida students and planned to establish contacts and exchanges with Naxi school children.[103] The Simon Fraser group leaders were so enthusiastic to learn about the prospects for Jade Dragon tourist trekking that they offered to help train our villager friends from Wenhai and Yuhu to manage the planned business.[104]

At Governor Xie Ye's request, I prepared written notes and gave advice to assist with his efforts to obtain World Heritage Site status for Dayan, the Old Town sector of Lijiang. We also urged him and his advisory council to consider applying for World Heritage status for the Tiger Leap Gorge and its surrounding mountains.

I departed, this time from a very modern Lijiang airport, opened only a few weeks earlier, to attend an international conference in Chiang Mai, Thailand. Seth and Lindsey continued their fieldwork for an additional month. The conference in Chiang Mai was entitled "Montane Mainland Southeast Asia in Transition." I presented the opening paper on our Lijiang work, written jointly with Professor He. The four days of papers and discussions (12–16 November 1995) appeared highly worthwhile with many valuable political-developmental overtones. The proceedings were published the following year (Rerkasem 1996).

Fig. 116. The story of the Love-Suicide Meadow as described on an ancient Naxi pictograph. This form of written language is claimed by the Naxi to have been developed before the Chinese invention of writing.

Fig. 117. Portrait of a young Yi matron willing to momentarily turn her attention away from her finely arrayed horse and potential customers for the ride to the Love-Suicide Meadow.

Once back in California, I prepared a detailed report on work accomplished together with a request for a three-year extension of the original grant. About the time this was ready to be submitted, on 3 February 1996, Lijiang suffered a high-magnitude earthquake of 7.0 on the Richter scale. The epicentre lay beneath the Tiger Leap Gorge. The damage was widespread and severe. Throughout the region, many buildings had collapsed and there was considerable loss of life. Our farm house at Wenhai that was planned as a trekking lodge was badly damaged. Much of the Old Town of Lijiang was heavily disrupted and many buildings were destroyed. Ironically a group of recently built incongruous concrete structures remained intact. Governor Xie Ye and I had discussed earlier that they would have to be dismantled if World Heritage status was to be granted.

In my letters of condolence to the governors of Yunnan Province and Lijiang County, I asked that, notwithstanding the devastating situation, the concrete structures encroaching on the lovely Old Town remain on the list of condemned buildings if at all possible. I also telephoned my senior contact at the World Bank with a plea for financial support to assist in the recovery from the earthquake. He responded that he would do everything in his power to help. I learned much later that the World Bank had indeed provided considerable assistance.

The modern concrete buildings were demolished; the Old Town was restored after the earthquake; World Heritage Site status was confirmed by UNESCO in 1997; but our dream of trekking the Tiger Leap Gorge and the Jade Dragon mountains remained a dream. Our request to the Ford Foundation for a three-year extension to enable us to take advantage of the progress made to date was rejected, despite our having won strong support from the villagers and all levels of government in Yunnan.

Lijiang and Northwestern Yunnan Development after 1996

A few years after our 1995 visit, IUCN, on behalf of UNESCO, asked if I would review a lengthy proposal submitted by the Chinese government for designation of a large section of northwestern Yunnan as a natural World Heritage Site. This region had already

Fig. 118.   "Waiting for clients" – Yi tourist business headquarters off the main road north from Lijiang and at the base of the trail to the Love-Suicide Meadow. The high peaks of the Yulong Xueshan form the skyline.

been declared the Three Parallel Rivers National Park. The area included three of the world's most spectacular river gorges; also, it was inhabited by a large number of minority peoples, an unusual situation for natural site nominations. However, the Haba Xueshan and the Yulong Xueshan, twin sentinels above the Tiger Leap Gorge, despite being close to the park's eastern boundary, were not included. I prepared an extensive review that was basically highly supportive. However, I included two critical requirements: first, it was imperative that a much more detailed plan for the welfare of the minority people within the proposed site be included; second, the northeastern perimeter should be extended to embrace the Yulong Xueshan and, hence, the gorge and the Haba Xueshan.

The combined report, with input from many reviewers, was submitted to Beijing for comment, as is the normal course for this kind of procedure. Some months later, I was sent the adjusted proposal and found it highly satisfactory. World Heritage status was approved by UNESCO in July 2003. It seemed that it was a highly worthwhile outcome,

Fig. 119 [top].   We were told that in the time of Joseph Rock's long sojourn trout could be caught from the front gate of many of the houses in the centre of Lijiang. This view shows typical "downtown" Lijiang (Dayan) in the early 1990s.

Fig. 120 [bottom].   One of the main channels of the river on the edge of Lijiang's still traditional market place where in 1985 the pelts of the near-extinct pheasant could be purchased for the equivalent of less than a dollar.

Fig. 121. Portrait of "the laughing monk" taken one afternoon in the quiet of his monastery garden, the result of a long convivial but mutually unintelligible conversation.

assisted in part by the efforts of our team. But was that really the case?

Shortly thereafter, I learned from Kunming that, in defiance of World Heritage regulations, plans were being prepared for a hydroelectric power station to exploit the immense energy potential of the Tiger Leap Gorge. However, there must have been many more objections than my own because the planned construction was halted, or at least delayed. So what has become of the Gorge, this incomparable natural treasure?

Lijiang is now one of the main focal points for tourists destined to visit the Gorge as well as a major tourist destination in its own right. The walking trail we first used on the western side of the river has been obliterated (see Figures 107, 108) and a road built so that tour buses can rumble through. Adventurous bus riders can walk down an extremely dangerous track, with inadequate guard rails, to view the rock from which the proverbial tiger leapt to safety from the hunters. This "development" is the result of the newly named Shangri-La County (until 2001 known as Zhongdian) which competed with Lijiang County to attract the maximum number of tourists. The competition arose because the boundary between the two counties follows the course of the river along the bottom of the gorge. Lijiang County responded by blasting a road part way into the gorge on its side of the county line to an overlook so that, with a minimum of discomfort, tourists can gaze down on the site of the mythical tiger leap. Blasting and road construction on both sides have destabilized sections of the canyon walls, while rockfalls and landslides have not only caused extensive environmental damage but have repeatedly disrupted transport.

In Lijiang's Old Town, "development" over the last 15 years is hard to assess, at least from afar. Consequently, the following remarks depend on first-hand hearsay. From photographs I have been shown, the World Heritage Site has become a spectacular tourist mecca. Trout have returned to its once "sweet water," now clean again. The buildings are alight at night; fine restaurants flourish; antique shops are full of valuable objects; the town is teaming with visitors. Most of the original Naxi inhabitants have sold their property to rich Han Chinese entrepreneurs and moved away. What has happened defies World Heritage Site regulations. But compared with Kathmandu, dirty, piled high with garbage, with open sewers, beset with grid-locked traffic and its accompanying air pollution that on many days conceals the mountains, a disease-laden trap for tourists and inhabitants alike, Lijiang (or Dayan) is a jewel.

Yet, it remains to speculate whether the thousands of minority people, especially the villagers distant from the "big city," are any better off today than they were on my first visit in 1982. We would need another research grant to provide a reliable answer; I am afraid to guess.

Fig. 122.   Shanzidou (highest point of the Jade Dragon
Snow Mountains) and its neighbouring summits after a fall
of snow, November, 1995.

Fig. 123. The High Pamir, Tajikistan, from which glaciers debouch that are as much as 65 km in length (1987).

CHAPTER 15

Roof of the World: The Pamir, 1999
"A Catastrophe of Biblical Proportions"

Perhaps of all the world's mountains, the Pamir (Roof of the World) hold pride of place for mystery, romance, inaccessibility, and beauty. Sometimes referred to as "The Pamir Knot" on a world map, this great complex of high mountains appears to tie together the Himalaya, Karakorum, Hindu Kush, Tian Shan, and the Kun Lun Shan. Marco Polo and his entrepreneurial uncles passed beneath their shining summits en route to the court of Kublai Khan; the "Great Game" confrontations of the Russian and British empires frequently played up against their flanks; the mighty river Amu Darya and its headstreams, fed by melting snows and glaciers, rolls down through precipitous gorges toward Samarkand and into the flatlands of the Karakum desert where Sohrab and Rustum fought their tragic day-long duel.[105] Today, modern and much more prosaic yet stirring tales arise from the High Pamir. What follows is one of the more modest of such tales, although the widely accepted speculation of a catastrophe that would threaten millions of lives is certainly far from prosaic.

The Aga Khan Foundation's U.S. affiliate, Focus Humanitarian Assistance, invited Don Alford to arrange for a discussion to outline a practical aid programme with emphasis on natural hazard abatement for Northern Pakistan, Gorno-Badakhshan (eastern Tajikistan), and northeastern Afghanistan. It was to be held in Salt Lake City in June 1996, on the campus of Utah State University, where Don held a professorship at the time. In addition to Don, the meeting was attended by a group of Ismailis, led by Behrooz Ross-Sheriff, Steve Cunha, who had recently completed his doctorate based on fieldwork in the Pamir, Pauline, and myself.

What I remember most from this meeting was Steve Cunha's urging that Lake Sarez in the High Pamir be considered a possible serious threat if rumours about the instability of its landslide dam were to prove correct. If the dam collapsed, the communities living downstream far beyond the mountains and all the way to the Aral Sea would be at risk. He showed colour slides of the landslide-dammed Lake Sarez, more than 60 kilometres long, which he had visited in 1993 as a member of our UNU research team. He suggested that Lake Sarez should be investigated, if only to allay fears of a possible catastrophic dam collapse.

J. D. Ives, *Sustainable Mountain Development*, https://doi.org/10.1007/978-3-030-96029-2_15

Several years of indecision were to pass. In early March 1999, I greeted Don Alford's telephoned "call to duty" with scepticism. He asked if I would be a member of a UN/World Bank/Focus Humanitarian Assistance team to reconnoitre the potential hazard of Lake Sarez. I agreed to join the proposed mission but remained doubtful that it would materialize because of the lapse of time since our meeting in Salt Lake City. Then, later, when Alessandro Palmieri of the World Bank phoned to ask if I would take the role of leader of the socio-economic unit of the proposed interdisciplinary and international team, I readily accepted.

Alessandro asked me to propose a couple of geomorphologists with mountain hazards experience who might be willing to accompany us. I proposed Steve Cunha and Teiji Watanabe, both former doctoral students of mine, as indeed was Don Alford. I was told that a Japanese was to be preferred over an American by the UN/World Bank combination that was financing the venture, a disappointment in the making for Steve. Next I was asked if I would be a member of the World Bank component, but *pro bono*, a surprise although the lure of the Pamir prevailed. Finally, there were practical questions and urgings for me to make my own air reservations. Kazak and Tajik visas would be required as our route to Dushanbe, the capital of Tajikistan, would be via Geneva, Frankfurt, and Almaty in Kazakhstan in order to avoid Moscow and Uzbekistan, on political considerations. It now appeared that this mission to the High Pamir would actually happen. I left Ottawa, en route for Geneva via Heathrow on 30 May 1999.

I joined the other expatriate members of the team at the Palais des Nations in Geneva. Scott Weber, the UN International Decade Natural Disaster Relief (IDNDR) leader, a young American, introduced us to each other: Jörg Hanisch (Germany), Carl-Olof Soder (Sweden), Gerard Leclaire (UK-Channel Islands), Alessandro Palmieri (Italy and World Bank), Bruno Periotto (Italy), Attilio Zaninetti (Italy), Teiji, Don, and me. We selected abundant supplies of camping equipment from the UN stores in the Palais des Nations – enough also for the Tajik members who would join us in Dushanbe and Khorog. We also assembled an assortment of food as we did not want to impose on our Pamiri hosts who would include many struggling subsistence farmers. We made sure that there was an abundance of Swiss chocolate and Nestlé condensed milk in tubes.

We travelled to Dushanbe via Frankfurt and Almaty as originally planned. Two days after leaving Geneva, we were met at Dushanbe airport by the deputy prime minister of Tajikistan, H.E. Ismat Eshmirzoev, and various UN officials. This ensured that we were excused customs and passport inspection, and a fleet of blue-flagged UN vehicles quickly conveyed us to the UN Development Programme (UNDP) building in the centre of the city. Here we were welcomed by the UN deputy resident representative, Blanche de Bonneval. This somewhat imperious lady in her elegantly redecorated office read us the rules of the forthcoming enterprise. Dushanbe was under military curfew; if any of us were taken into custody between 10 p.m. and 6 a.m., we would be immediately sent home. Extreme care would be essential, especially since a UN peacekeeping force was maintaining a fragile balance between central government forces and opposition militia. Some militia units had not joined the UN-negotiated temporary accord and about 25% of the country's territory was out-of-bounds and classed as dangerous (four UN personnel had been ambushed and assassinated three months previously).

We met our Tajik counterparts, military and scientific, and our two UN team members, both women, one of whom turned out to be a liability, the other, Lise Grande, one of the most competent and courageous persons I have met.

Our complicated logistics were worked out – a convoy of four UN trucks would proceed by road with most of our food and equipment and join up with the Focus Humanitarian Assistance vehicles in Khorog; a military helicopter would fly in the geologists, geophysicists, and engineers who were to stay at the lake, plus a few others, including me, who would visit the lake for the day and then return to Dushanbe with the helicopter. Together with the rest of the party, we were to fly back into the Pamir the following day by twin-engine turboprop and so reach Khorog, the capital of Gorno-Badakhshan. Thus we were split into two groups, one based at Lake Sarez, the other to work out from Khorog in the deep valleys below the lake.

Next came a mass of telecommunications equipment, including satellite phones and walkie-talkies so that each vehicle could have constant contact with Khorog, the UNDP in Dushanbe, and with each other. We all received detailed operating instructions so that each of us could use the equipment independently. Finally, something of a shock! Release forms were distributed for our signatures. The UN and the World Bank would assume no responsibility in the event of accident, death, illness, or forced evacuation. I was greatly disturbed (as were all members of the team) on a matter of principle and organization. Without our signature, we would not be allowed to continue. Had I been twenty years younger with dependent children, I would have immediately opted to return home. Perhaps this was the reason why we had been told at the outset that we would be operating *pro bono* and without contracts. We all felt that we had been improperly treated, although the lure of the High Pamir won the day.

Lake Sarez, the Dam, and the Mission

In 1911, the central Pamir experienced a major earthquake, estimated at above 8.0 on the Richter scale, although seismic instrumentation was very sparse and rudimentary in those days. The earthquake caused a gigantic landslide, estimated from the deposits to be in the order of three to four cubic km in volume. In effect, an entire mountainside fell into the Murghab River valley and dammed it to a height of nearly 600 metres. The Murghab River, a tributary of the Bartang River, which flows for about 130 km before joining the Pianj, itself a major tributary of the Amu Darya, backed up behind the dam to form Lake Sarez. The dam was named after the village of Usoi, which it had obliterated, and the lake was named for the village of Sarez, which had been submerged in 1912 by the rising waters. In 1999, the lake surface stood at 3,200 m above sea level; it was continuing to rise at about 5 cm/year and the lowest part of the dam crest was still 50 m above lake level. Lake Sarez drains internally through the very broad dam and flows down its original gorge to join the Bartang River. The lake is over 60 km in length and contains about 17 km³ of water, comparable in volume to more than half that of Lake Geneva.

The setting of Lake Sarez would rank it as one of the most spectacular mountain lakes in the world; this is accentuated further by its colour – an unbelievably deep metallic blue. It is surrounded by high, snow-capped and glacierized mountains, the highest exceeding 6000 m.

Long stretches of lakeshore are overlooked by extremely unstable slopes from which small landslides and rockfalls frequently release into the lake. These massive falls of rock and gravel generate powerful waves, the height of the wave crests being proportional to the volume of the landslide and the impact of hitting the lake. There is no record of these waves coming remotely close to overtopping the dam. However, if a very large landslide were to release, this could cause a wave to overtop the dam and so endanger the lives of everyone living in the Bartang Valley (approximately 5000 people) and perhaps far beyond. This geomorphic instability is further accentuated by the occurrence of earthquakes; small quakes occur very frequently and also release landslides. The Pamir form one of the most seismically active regions in the world (see Map 9).

The Murghab-Bartang river extends for more than 130 km before joining the Pianj River, which forms the international boundary between Tajikistan and Afghanistan. Khorog, a town of about 10 000 people (in 1999), lies on another tributary of the Pianj about 80 km upstream of the Bartang-Pianj confluence. The Bartang Valley is a tortuous mountain gorge. The road along the bottom is rudimentary and is subject to landslides, mudflows, floods, rockfalls, and avalanches. The climate of the valley bottom is extremely arid, although high peaks with snow and glaciers, as well as the main river, where accessible, provide irrigation water to the 30 or more small villages (locally called *kishlaks*) that exploit every available square metre of useable land. Downstream from the Bartang-Pianj confluence, the Pianj flows through a deep gorge, enclosed in places by near vertical walls, except for short flat areas where tributary streams have cut their way through and laid down steep alluvial fans, dangerous sites for the small villages on either side of the international frontier.

Soviet scientists had monitored Lake Sarez for more than 50 years, but their electronic early warning system would have alerted only Moscow and Dushanbe. Thus, if the dam had collapsed for any reason, the approximately 5000 Pamiris in the Bartang Valley, together with an unknown number in the Pianj Valley, would have been swept away. In the event of a dam failure, it would take only 17 minutes for the deluge to reach the nearest village!

The Soviet automatic sensing system at Lake Sarez had fallen into disuse following

Map 9. Outline map of Tajikistan showing locations of Dushanbe and the Pamir Mountains, Lake Sarez, the Pianj River, and Khorog.

collapse of the USSR in 1991. In recent years, there has been growing speculation that a total failure of the dam would not only have disastrous consequences, but also that failure may be imminent. The reason for our team to be in Tajikistan in June 1999 was in response to widespread alarms voiced by the governments of the new republics of former Soviet Central Asia, that total collapse of the Usoi dam was likely. The so-called worst case scenario anticipated the discharge of 17 km³ of water, projected to pick up an equal volume of debris and rush down the Bartang at speeds in excess of 100 kph, sweeping through the Pianj valley and reaching as far as the Aral Sea. Such an event would be a natural catastrophe

of unimaginable magnitude and affect the lives of five million people in four different countries. However, this worst case scenario had been modelled by the U.S. Army Corps of Engineers and was based on the assumption of total failure of the dam, an entirely hypothetical construct, not a prediction. Unfortunately, politicians and the news media have a tendency to confuse hypothetical constructs with predictions.

Our task was to ascertain, as accurately as possible, the degree of danger and the possible damage that would occur in a number of different scenarios. We also were asked to determine the attitudes of the local Pamiri who would be the most vulnerable in the

Fig. 124.   The western Pamir provide a mountain landscape in stark contrast to the High Pamir (Fig. 123), Tajikistan.

event of any catastrophe. This task was to be my primary responsibility and for the anticipated numerous interviews of local people I was to be assisted by Goulsara Pulatova, a highly educated Pamiri who was employed by the Aga Khan Foundation. Svetlana Vinnichenko, director of the Tajik Geological Survey was another member of our team who joined us in Dushanbe. After completing the field investigations, we were to make formal recommendations to senior Tajik government officials, to be followed in Geneva by an international press conference. The final task would be a peer-reviewed publication that would include an appraisal of the large volume of information resulting from earlier Soviet investigations.

The multi-faceted mission was made more sensitive by the existing political situation and the degree of inaccessibility. North-eastern Afghanistan, including the Wakhan Corridor, was still struggling to maintain

Fig. 125.   The geologists and engineers are setting up base camp above the Usoi dam. Lake Sarez is largely hidden by the helicopter. Colonel Usmanov, partially uniformed, is on the left with Lisa Grande, conspicuous in yellow, and other members of the team (1999).

Fig. 126.   Before leaving our homes, the expatriate members of the survey team were told that the colour of Lake Sarez would impress us.

Fig. 127.   The military helicopter flew us over Lake Sarez and down the full length of the Bartang Gorge. This view provides a perfect example of an alluvial cone on which every square metre of usable space is occupied by one of the 32 small villages (kishlaks).

Fig. 128.   Settlements in the Bartang Gorge are faced with severe hazards on a day-to-day basis. These are much more real to the Pamiri inhabitants than the supposed threat of a landslide-dammed lake that they have lived beneath for a hundred years.

its independence of the Taliban forces who were seeking complete control at that time. The people of northeastern Afghanistan were suffering from the after-effects of a series of disastrous earthquakes of the year before. Northern Pakistan, China, and the now-independent republics of the former USSR were all irretrievably part of the larger geographic setting (Map 9). Our team included engineers, geologists, geomorphologists, hydrologists, social scientists, computer (GIS) experts, and senior Tajik military, as well as staff of the Focus Humanitarian Assistance group, the UN personnel and, perhaps most important of all, our drivers.

The helicopter flight to Lake Sarez to set up an advance camp for the geologists and engineers was an excellent opportunity to photograph the Pamir in good weather, and especially the startling blue of the lake. On the return to Dushanbe a detour down the Bartang Gorge afforded an outstanding view of the 600-metre high landslide dam that held back the immense body of water that appeared to be threatening the many tiny villages strung out along the floor of the gorge.

The next morning's fixed-wing flight from Dushanbe was equally spectacular, across a wide swath of the Afghan Hindu Kush, down through the Rushan Gate, wing-tips seemingly close to the rock walls, along the Pianj River and so into Khorog. Logistical problems, not unexpected, cropped up immediately. The first attempt by the UN four-truck convoy to reach Khorog with our equipment and food had been turned back from the "direct" route by reported heavy rockfalls in the Pianj Gorge. The alternative route took the convoy north

Fig. 129. In places the Bartang Gorge presents a challenge to route construction engineers. In this section early summer snowmelt would quickly raise the level of the river and render the upper gorge inaccessible.

from Dushanbe through short sections of Uzbekistan and Kyrgyzstan, across the Alai range and so around the Pamir to approach Khorog from the east. It involved two very long days and a hypoxic night on the Pamir Plateau above 4000 m – which, we were later told, had caused nosebleeds and considerable respiratory discomfort.

"If Allah Wills the Flood, I Will Be Ready," or What the Partridge Says

For three days, with improvised transport while we awaited the diverted convoy, we drove out from Khorog in small groups in different directions to familiarize ourselves with the region. There was one long day down the Pianj Valley on a very good road when I made many house calls to interview the local Pamiri and ascertain their attitudes toward the threat of Lake Sarez. Goulsara provided the essential interpretation. Many of the villagers knew her from previous visits, which assured us of a warm welcome and ease of interaction. It quickly became apparent that they had little

Fig. 130. "If Allah wills the flood, I will be ready." This wonderful lady would have killed one of her precious chickens to ensure traditional Pamiri hospitality had we not stopped her by describing ourselves as the bearers of UN food aid (1999).

concern about the possibility of a catastrophic flood. They appeared more worried that they might be forced to leave their homes because of overreaction by the central government.

In the village of Shipad, the farthest downstream that we reached, we met a kindly old lady who had lived in her tiny house all her life. In fact, she had been born there, and her parents had been living there in 1911 when the giant earthquake and landslide had occurred – they had been unaware of any disaster at the time until they were informed several days later by word-of-mouth! Her attitude was that, as an old widow, she would always remain in her house: "If Allah decides to cause a flood, I will be ready for it, but I don't think He will."

She was very poor. Her possessions amounted to her house, two walnut trees, six mulberry trees, and several chickens. She made a move to grab and kill one of the chickens so that she could provide us with lunch and scolded Goulsara when we stopped her. The risk of affronting local hospitality customs was averted by our showing her what we were carrying in our truck. We produced all necessities for lunch, except the tea, which was certainly the preserve of our hostess, and we gave her an ample share of the UN provisions. Next, when I showed her a coloured photograph of my daughter and three red-haired grandchildren, she was jubilant, insisting that I was such a young man to be so richly blessed. On

Fig. 131. Our Pamiri host family after an evening of feast and dance. Basid village in the Bartang Gorge below Lake Sarez (1999).

discretely ascertaining her age, I amazed her by pointing out that we had both been born in the same year. On our departure, she presented me with a beautiful pair of traditional Pamiri socks that she had knitted herself.

Our colleagues at the lake included Jörg and Carl, the geologists, who were to work there throughout the length of our mission, to be picked up on the final helicopter sortie. They were accompanied by the Italian engineers Alessandro, Bruno, and Attilio, who were to walk out down the Murghab Gorge and be met by local guides who would travel by jeep as far up the Bartang as possible and proceed on foot to meet them. Our traverse of the Bartang was planned to coincide with the walk out by the Italians. It would allow visits to several of the Bartang villages, spending one night in the upper valley and returning the next day to Khorog.

Leaving Don and Behrooz in Khorog, we set out early on 10 June for the Bartang Gorge in the four UNMOT vehicles. Hiralo was my driver. Teiji, Gerry, Svetlana, Goulsara, and Lise followed in the remaining three, each with an official driver. After a fast two-hour drive, we passed through Rushan and entered the Bartang Gorge, meeting the sunburnt Italians about an hour later. Following a short exchange on road conditions and news of their work at the lake, we left them to return to Khorog and continued our tortuous journey upstream. We eventually split into two groups: I chose to stay with Gerry, Goulsara, and Lisa overnight in Basid, some 40 km short of the end of the road. The others proceeded to the confluence with the Murghab River below the landslide. Our overnight in Basid saved 40 km

Fig. 132.   Children the world over find strange visitors objects of great curiosity. These Pamiri village children in the Bartang Gorge are no exception.

of travel the next day, allowing more time for interviewing villagers and photography on the way down.

We were received with enthusiasm in Basid, stayed in the house of a very hospitable Pamiri family who feasted us and then invited us to dance with them until the early hours. It was an enlightening experience. The Western guests were somewhat hesitant about proper procedure. It was the laughing, friendly Muslim women who took the initiative and literally dragged us out to dance with them. They were Ismaili Muslim and very different from their Sunni and Shia co-religionists. Afterwards, we slept on the floor with the traditional Pamiri rhythms forming accompaniment to our dreams.

Amongst the sound and confusion of the evening, Hiralo unobtrusively obtained his long-sought Pamiri partridge, the personal objective of his involvement in our mission and the explanation for its inclusion in the heading for this section of the chapter. The people of Basid were well known for raising partridges that were taken as chicks from their nests and reared in captivity. They were in demand over quite a large territory and used in the cruel "sport" of fighting, similar to earlier Western cock fighting. For years, Hiralo had longed for one as a pet. And the next morning, the young adult partridge was snugly housed in a wicker basket behind me on the seat of our truck.

Following a generous breakfast and several group photographs, we set out downstream to retrace our road to Khorog, stopping frequently to visit villages and take photographs. It was an enjoyable day, long and hard, with generally excellent weather. Hiralo, very pleased with himself, nursed his prized partridge which occasionally made low comforting partridge sounds despite the incessant jolting of the truck.

The two days in the Bartang Gorge reaffirmed my earlier findings about the local attitude towards the supposed threat of Lake Sarez. It was apparent that there was almost universal disregard for the potential dangers and much greater concern about food availability now that Soviet subsidies had been eliminated, or about the multitude of actual mountain hazards that they had to face on a day-to-day basis. Yet, despite their seemingly precarious position, they wanted to stay put in their homes and were opposed to any possible forced evacuation.

Hazards of Mountain and Military Obstacles

After dinner in Khorog we were harangued by Lise, our UN group leader, to help work out the logistics for the next day. Despite her earnestness, in the midst of the deliberations, we were disrupted by wild cries – one of the drivers, curious about the partridge, had inadvertently let it loose. The bird was perched on the roof of one of the compound's warehouses as an anxious Hiralo crept up on it. As he tried to grab the bird, it slipped through his fingers and flew into town – despair!

Resuming our discussions with Lise, it became clear that there was a conflict of opinion about the choice of the route we should take for our return to Dushanbe. We had the choice of waiting for the helicopter and hoping it would be able to land at Rushan; of taking the long haul by truck across the Pamir Plateau and through parts of the neighbouring republics; or thirdly, by driving the direct route down the Pianj Gorge. Colonel Usmanov, head of the State Committee for Emergencies and our senior military member, advised that the Pianj road, while difficult, was as open as could be expected, and he recommended it as a reasonable route.

It was decided eventually that Teiji, Gerry, Svetlana, Goulsara, Lise, and I would go with the four UN trucks via the Pianj Gorge. Don, Berhooz, and the Italians, all of whom had more work to do locally, would go by helicopter and worry later if it was not able to land. The geologists Jörg and Carl-Olof and their assistants were to be retrieved by helicopter directly from Lake Sarez. The Pianj River party would make an early start in the morning in order to ensure that we were out of the gorge before nightfall, since rockfalls could occur at any time and the longer we stayed by the river, the higher the risk.

At breakfast, we were met by a jubilant Hiralo with partridge in hand. He had wandered into town and found that a local Pamiri had managed to catch the bird, so that, for a modest reward, it was back in its wicker cage and ready to go with us down the ferocious Pianj Gorge. Col. Usmanov had been in radio contact with members of his staff along our route and assured us at breakfast that no rocks had fallen during the night. He felt relaxed about the Pianj route, provided we were extremely careful and did not get cut off by nightfall.

Our four-truck UNMOT convoy was loaded, UN flags properly primed, and goodbyes exchanged. I was in the first truck with Hiralo, as driver, and the partridge as backseat passenger carefully swaddled against the anticipated rough ride. The trucks had radio contact with each other and with Khorog and Dushanbe. We had tested the satellite phones previously, but because we were almost always deep within mountain gorges, we could only make outside contact when one of the communications satellites was almost overhead. This prompted me to make an experimental

call home to Ottawa. The connection was so clear that Pauline thought I had arrived home early and was calling from the airport for her to pick me up in the car.

We left Khorog early on 12 June and made rapid progress to the confluence with the Pianj River. Once in the main valley, we speeded up to 60–70 kph. However, we had to grind to a halt on a sharp bend at the first military checkpoint. Young Tajik soldiers in Russian uniform and equipped with Kalashnikovs sauntered towards us waving and smiling. We jumped out and embraced them and, arm-in-arm, posed for photographs.

We rapidly passed through Rushan, with a last glance up the Bartang Valley, and shortly after reached Shipad. After Shipad, we entered a tight gorge section. Here there was little to encourage settlement on the Tajik side on account of the soaring canyon walls surmounted by imposing turrets of sheer rock. In places, the main river was such a narrow roaring white torrent that a cobble could have been thrown across to bounce off the far rock wall.

Our speed progressively slackened as the condition of the road deteriorated. Old mudflows and rockfalls, crudely bulldozed, reduced speed to a crawl. But within 45 minutes, the gorge opened out and more scenic tributary valleys came into view. There were many *kishlaks* on the far side of the Pianj, each connected by a distinct trail: sometimes far above the river to circumvent stretches of near-vertical slopes that plunged directly into the water; sometimes the route was almost in the river itself. Long lines of irrigation canals sneaked down side valleys toward the houses and, on reaching the lower, gentler slopes, burst forth into bright green fields with a variety of crops that contrasted vividly with the harsh browns backed by the high snows of the more than 5000-metre peaks deep into the Hindu Kush.

By 1:30 p.m. we entered a sizeable village and stopped for a hasty lunch in a commodious, if gloomy, restaurant. Little time was spent eating. There was a tense dispute between the drivers, Lise, and several village "experts." It appeared that there was trouble ahead, or at least uncertainty which, in our present situation, amounted to the same thing. The dispute heightened: should we continue as fast as possible, or stay in the village overnight and cross the "pass" early the next day? The climax of the drive out from Khorog, usually referred to as the pass, was estimated variously to be from three to six hours downstream. Could we make it before dark? There was no question of spending a night downstream in the gorge on account of the serious hazard of rockfall and debris flow, nor could the drivers continue after sunset. Finally, the discussion reached a crescendo as Lise harangued the quibbling disputants. She told them that we would not waste any more time in ill-informed argument but would "bloody well get going." Thus ended our 30-minute lunch break, the only respite the drivers had until many hours after sunset when, bleary-eyed, we reached our target, a UNDP encampment. Despite the appearance of bad feelings, smiles reappeared beneath the wand of Lise's firm hand, and we set off again towards the pass.

The road, nonetheless, continued to be hazardous – fords across tributary streams, rockslides minimally cleared. Within less than an hour from our lunch break, however, we met a jeep coming towards us whose driver told us that we were only 90 minutes from the pass. It was then only 2:50 p.m., so it appeared that we had plenty of time. That, however, was not quite accurate.

Shortly after the encounter with the jeep, the road narrowed and had been cut by bulldozers across a formidably steep slope that swept down directly to the river. A sizeable mountain stream fell vertically for 20 to 25 metres to hit the road with a creditable splash. It was obvious that we would have to drive on through and that our outer wheels would be centimetres from the drop-off into the river. Hiralo and I, in the leading truck, were to make the first try. We inched forward until the waterfall was consuming the hood and windshield and the vehicle was shaking in an alarming fashion. At this point, we were blinded. Hiralo, without sight and in a very tight space, rolled another three or four metres. We could see again! The waterfall was now pounding our cargo behind the cab. We squeezed through to the far side as our colleagues looked on with apprehension. During the pause, our partridge fluttered and cooed; Hiralo smiled, drove gently forward for a hundred metres to a reasonably safe spot, sat back, and awaited the others to make the passage.

Ahead of us the track rose steeply, levelled out, and then rose again. A rockfall had blocked the road close to the top of the second incline, about 200 m above the Pianj River. On the level section, a dilapidated bus was moving forward very slowly as several jeeps inched along behind it. It began to ascend the final incline. Halfway up, it stopped. From our position about 600 m behind, we could see some 20-25 passengers climb out; we detected what we assumed to be a dispute. Eventually the passengers began to walk back down the road, and the five jeeps painstakingly backed down, with many pauses as they repeatedly came rather close to the edge. Next, the bus began to back down, even more slowly. It reached the level section and stopped on a slight curve. From the top, a bulldozer began to move slowly toward the bus, which appeared to have become stuck. Several of us walked up to the bus, which was close to a short wider section where two vehicles could overtake. With a little care, it could surely back up another 10 to 15 metres and so allow the jeeps and our trucks to pass. Yet this was not the main complication.

Some time previously, a rockfall had hit the road close to the top of the second incline. A jeep had been knocked off into the river and, together with its occupants, had submerged. Apparently there were no survivors; a search had been ruled out as dangerous and a waste of time. A road crew with a second bulldozer, still beyond our line of sight, was clearing the road of the rockfall debris; a drill crew had begun to drill horizontally into the rock wall just above the road with the intention of loading the drill holes with dynamite and blasting to widen the damaged section. As a geomorphologist I was appalled. I glanced at Teiji and could sense from his expression that the same thoughts were coursing through his mind. It was not beyond possibility that dynamiting could cause a serious rockfall.

Perhaps we could persuade the drill crew to let us through before they decided to have their fireworks display. The bus driver seemed reluctant to back any further but attempted to ascend again; only halfway up, he stalled. Well, I thought, not a good place to have to camp overnight.

Everyone seemed compelled to give advice to the bus driver, much of it probably contradictory. Eventually Hiralo tinkered with the engine; it spluttered and started. Now the driver would surely back down to the place where we could all pass him. Instead, he continued upwards – perhaps he also had had a university course in mountain geomorphology.

Feebly, and with much backfiring, the bus trundled up the steep slope toward the crest. We held our collective breath. If he stalled again, we would never get through. But he continued. And as the distance between us and the bus extended, the noise from the engine that reached our ears lessened so that the roar of the river below us grew loader and seemed to enhance the precariousness of our position. The bus eventually disappeared over the crest and the waiting jeeps and our UNMOT trucks followed. Most of us chose to walk behind.

The crest of the road was a shambles of bulldozed boulders and gravel, with scattered large rocks too heavy to move. We pleaded with the drill crew to delay any detonation until we had passed and to withdraw the drill rods that were blocking our way. With some persuasion they agreed.

We waited patiently. Small pebbles skittered off the precipices above us, seemingly to warn us not to stay too long. Eventually the drill rods were pulled out and there was space to pass. We crested the high point of the road, the bus allowing us to overtake it, and began a long descent to river level. We had crossed the much-discussed pass.

Now close to the river, we could see the gorge open out ahead. Large valleys entered from both sides. The sun had dropped behind the towering cliffs and soon twilight would descend. We drove through several *kishlaks*, including the one where our outward convoy had turned back to Dushanbe because of reported rockfall on the pass. The road remained atrocious, despite the much gentler terrain. We climbed out of the gorge as twilight enshrouded us and reached open rolling country with small scattered villages and isolated farms, by which time it was quite dark, starlit, with a splendid Milky Way above us.

Our convoy had become strung out over more than a kilometre. There were periodic bursts of staccato conversation on the car radios; Hiralo seemed to be giving the drivers behind us timely warnings as we hit bad spots in the road. Then we came to a sudden and complete stop as our headlights shone onto a formidable metal barrier. A group of fully armed, starlit figures rushed toward us. They demanded that Hiralo and I get out.

Hiralo displayed a variety of documents, including a *laissez-passer* signed by the prime minister. This only caused muffled laughter. Recalling our previous checkpoint encounters, I got out ready to face a friendly embrace. I was mistaken. In the darkness, the silence broken only by Hiralo's firm, but obviously angry voice, there was the ominous metallic snapping of ammunition clips being slotted into Kalashnikov rifles.

We re-entered the truck, and an armed soldier got in behind us. He pushed aside the poor partridge and rested the barrel of his Kalashnikov on the base of Hiralo's neck. It became chillingly obvious that we were to drive only where he directed. Meanwhile, our second truck caught us up and was treated in the same manner.

With the Kalashnikov at his back, Hiralo turned off the main road; we proceeded slowly under the soldier's direction into what appeared to me as a blank nowhere. Presently, as I began trying to communicate with Hiralo, I was startled to feel something hard and cold pushed against the back of my neck. Hiralo quietly reached out his hand to press my arm. I sensed his message – keep still, the soldier was only trying to spread the tension. Nevertheless, it was a feeling that I will long remember. Trucks 3 and 4 were soon close behind us, and we drove on, silently, for about 20 minutes. It turned out that we were being escorted to the local militia headquarters.

We came to a halt in front of a row of buildings, barely apparent in the dark, set back behind a metal fence; there were more armed soldiers. The other three trucks drew up behind us. Hiralo and I had got out to face the soldiers when Lise walked up and proceeded to upbraid Hiralo and me, mainly Hiralo, for allowing an armed soldier to enter a UN vehicle – a distinct breech of international diplomacy (she could have been on Salisbury Plain, England: "not cricket, old chap!"). She then turned imperiously on the soldiers and demanded to see their commanding officer. At that moment, the chief officer emerged from one of the buildings surrounded by armed men and came up to us. I was very much impressed as, from my position, it looked as if Lise turned the commanding officer around and marched him back to his office as she called on us over her shoulder to wait and stay quiet. The dark air was heavy with suspense, the stars unwavering. We were ten tense captives, with four clearly marked UN trucks surrounded by about 20 men in military fatigues, all armed. There was a quiet murmur of voices; the black outline of hills in the distance could just be distinguished against the lighter starry sky. I was not surprised to sense the anxiety of my colleagues, as well as my own.

After about fifteen minutes (it seemed an age), a tight-faced Lise emerged: "Everything is settled; let's go." So we about-turned and with a feeling of considerable relief drove back to the main road. Much later, I asked Lise what she had expected us to do when an armed soldier forcefully entered our truck. She explained that it was standard UN procedure to express outrage and direct it at the home team rather than the actual offender – that way it was possible to get the message across and keep to a minimum the risk of an unpleasant response from a trigger-happy thug.

An hour later, we reached a small town and, more relevant to our now-pressing needs, a UNDP establishment. It was just before midnight. Bone-weary from 16 hours of jostling on the road and, I suppose, slackening mentally as the tension eased, we rapidly signed in, took hot showers, and sacked out.

I noticed Hiralo creep gently into his allotted room carrying the wicker cage draped with a cloth covering. No doubt the precious partridge was already fast asleep.

New Day, New Horizon, New Kind of Problem

The next morning unfolded with unbounded luxury. After a solid night's sleep, we breakfasted on rich coffee, bacon and eggs, marmalade and toast, fresh fruit. It was Lise's birthday, 36 we surmised, and we were ready to celebrate. We posed for group photographs in the still cool morning sun. The trucks were reloaded and we were off on the final leg of our journey to Dushanbe via Kurgan-Tyube, an estimated four-to-five hour drive.

Soon we were passing through semi-arid dry-farming country over now the "normal" bad roads. Occasionally vehicles passed us, usually trucks or a very occasional overcrowded bus. After little more than 30 minutes of travel, a sharp pain developed in my lower right abdomen. It became rapidly more acute until I was gasping with the shock. I signalled for Hiralo to alert the others and stop. This we did. Gerry and Lise ran up to our truck as I was gingerly eased out and prevented from collapsing only by alert drivers who grabbed my forearms. Painfully, I was edged to a fortuitously close concrete shelter and stretched out on its bench. Gerry, who had considerable paramedical experience, examined my abdomen and took my temperature. We worried about a possible ruptured appendix, but I had no fever. The practical Lise had one of

Fig. 133. The Pianj Gorge evacuation team the morning after, following a superb breakfast in celebration of Lise's birthday. From the left: Svetlana, Teiji, a driver, Lise, Gerry, Jack, and the other drivers and mechanic, including Hiralo, second from right (1999).

the satellite phones set up and was quickly in contact with UN headquarters in Dushanbe: what to do? helicopter evacuation? the hospital in Kurgan-Tyube? or what? It was decided to give UN headquarters a chance to summon a local doctor and to use a satellite phone for a remote examination. Could I wait another 45 minutes? If I could get back into the truck, could we proceed slowly towards Kurgan-Tyube so that its hospital would be that much closer? I decided to continue, so another sat-phone contact was arranged for 45 minutes later.

I felt an urgent need to pee! Hiralo and Gerry literally held me erect, the women circumspectly looking away. I couldn't produce. So I was half-carried back to the truck and we set off, slowly over a not too bumpy road. Regardless, the pain was still excruciating.

After 35 minutes we stopped. I was helped out again and inched along the road to where Gerry and Lise were setting up the sat-phone. I was propped up as we awaited contact. Presently, in accented but clear English, a male voice introduced himself as the local doctor. "Would Professor Ives please lower his trousers and underwear?" So there I stood, on the main road, "the full monty" in the blistering sun; perspiration from pain and the heat of the day (over 40°C) soaked my shirt. Lise and Goulsara discretely walked back to the rear of the trucks; a Tajik bus came into view, heading towards us, and swerved violently, nearly off the road, as the driver caught sight of the indelicate display. The voice over the sat-phone next asked Gerry to probe the obvious places – no increase in degree of pain? Confirmed. Next the doctor asked, "Please tell me, have the patient's testicles enlarged within the last 24 hours?" Gerry, somewhat non-plussed, does not know what to say and stammers that he is not accustomed to measuring male genitals, could I help him? There followed a roar of laughter.

The doctor suspected rectal colitis and suggested we drive back to Dushanbe if I could hold out, since it was advisable to avoid the hospital at Kurgan-Tyube. By this time, joking aside, I was well and truly sapped. I was gently bundled back into the truck along with Hiralo and the partridge, Gerry joining us with a supply of morphine. We proceeded.

Within 30 minutes or so, the intense pain amazingly began to subside. What had felt like a knife cutting into vital organs simply faded away. Within another 45 minutes, I could feel nothing excepting the general sense of exhaustion. I said I felt a fool to have caused such a fuss. Gerry kindly pointed out that if the only damage was to feel a fool, especially among friends, then we had the best possible resolution.

We reached Dushanbe just before 3 p.m. Steve Weber, our absent leader who had been back to Geneva while we were in the field, came from Dushanbe airport soon after with the news that the helicopter with the Lake Sarez group, Don, and Behrooz had landed.

We reassembled in the Hotel Tajikistan in Dushanbe and planned a birthday party for Lise. However, Lise had gone to her town apartment to refresh and must have been so completely exhausted that she slept all evening. Thus we partied without her, but retained half her birthday cake for the party we determined to hold the next evening when she would likely be with us.

The main debriefing was the following morning in the office of the deputy prime minister, Mr Ismat Eshmirzoev. The upper levels of the government had expected us to report that there was an imminent danger of the dam collapsing. This, of course, would have brought about the influx of very large amounts of international money. First, however, to open the meeting in a light-hearted manner, the deputy prime minister winked and said he was relieved that I was well and especially that I had not been arrested for indecent exposure on one of his country's main highways, something hard to explain away in a Muslim country: much laughter at my expense.

We then proceeded to the serious discussion about Lake Sarez. It became apparent that our unanimous refusal to adopt the worst case scenario – that a catastrophic flood was imminent – was a great disappointment. My more technical colleagues provided the reasoning behind their conclusions and answered a battery of questions. I gave a brief report on my interviews with the local inhabitants, explaining their most serious fear was that the government would panic and send in soldiers to forcibly evict them from their poverty-stricken but nevertheless, beloved homes and mountains. They were prepared to live with the danger of dam collapse, as the entire population had for more than 70 years, if an early warning system could be put in place and help given for them to combat the many lesser hazards that beset them in the lower Bartang Gorge. An extensive technical report was prepared and published the following year (Schuster and Alford 2000).

Despite the obvious disappointment, Ismat Eshmiroev graciously accepted our highly conservative recommendations and expressed his enthusiasm to continue to work with us in the future. After the meeting, I was invited to return to Dushanbe the following September to present the opening keynote paper for the mountain conference that was being planned, with all my expenses to be covered by the Tajik government.[106]

The deputy prime minister accompanied us to the airport the next morning and expedited our departure formalities. We thanked and said goodbye to our Tajik and UN colleagues. I worried how I could show my appreciation to Hiralo with a meaningful small gift. All I had was U.S. dollars, and the appearance of a tip would not be appropriate. The answer flashed into my mind. I told him I wished to buy from him that glorious partridge. Of course, not realizing my intent, he refused. But with a little more explanation, assuring him that it would not leave Dushanbe, he cautiously handed me the wicker cage. I gave him in return a hundred dollars, photographed bird and cage, and returned it to him saying that the precious partridge was now my present to him. Don and Teiji, standing nearby, were obviously mystified by this strange procedure. Hiralo set the cage down and we warmly embraced. "Come back to my country again," he said, "and I will drive you wherever you want to go." To which I replied, "Many thanks, Hiralo, you are the best driver I have ever known, I will try to take you up on your offer – and please call the partridge 'Jack'."

Back in Geneva, the entire team met to arrange a series of presentations that would be made the following day during a press conference set up by UN headquarters. Unfortunately, two inflammatory news commentaries had appeared immediately before our official reporting. Each cited its main source as Scott Weber of the UN Department of Humanitarian Affairs "who organized the expedition" [he had not accompanied us beyond Dushanbe on the way out, although he was official leader of the team]. Some of the more regrettable statements that appeared in the two news releases were these:

> "Scott Weber said . . . they [the research team] had found an enormous disaster waiting to happen";
>
> "Five million people could die";
>
> "When the natural dam which holds back the water breaks – which experts say could be at any moment – a wave as high as a tower block will blast a trail of destruction a thousand miles through the deserts and plains once crossed by the Silk Road"; and
>
> "We don't know when it could go, but it could go at any time."

In contrast, our press conference was attended by 20 prominent journalists, and their published reports reflected our far less dramatic assessments of the potential threat posed by Lake Sarez. Similarly, the government of Tajikistan accepted our findings, and work began to implement our recommendations soon after. Electronic monitoring and early warning systems were installed and linked to each of the 32 kishlaks in the Bartang

Gorge as well as Khorog and Dushanbe. Safe havens stocked with food and clothing were set up well above the level of any conceivable flood or debris flow. The Lake Sarez Risk Mitigation Project was formed with funding from Switzerland, the World Bank, and the UN. All finally seemed calm and rational. Then, in early April 2003, an alarm was sounded on a Russian website (www.strog.ru):

In Central Asia an accident on a planetary scale is expected. Today, Uzbek scientists have deciphered space images from the Japanese film-making system ASTER using the satellite TERRA. They discovered that Lake Sarez has over-topped the dam that is now being destroyed as if cut by a giant circular-saw.

The announcement referred to a 100-metre high mudflow that would destroy cities for 2000 km downstream to the Aral Sea, with 600 000 to 5 million lives lost (translated from the Russian by the U.S. Embassy in Dushanbe). Tense reaction reverberated throughout Central Asia and all the way to UN headquarters in New York and to Washington DC, as well as to members of the 1999 evaluation team. Sober, authoritative voices calmed the possibility of panic, although the post-1999 Lake Sarez Risk Mitigation Project planned to send a reconnaissance mission to the lake. No subsequent information reached me, however, and the very absence of news assured me that there had been no flood "of biblical proportions" with the loss of millions of lives, and that the April 2003 alarm was false.

Following the 2003 alarm, the level of Lake Sarez appears to have continued to rise about five centimetres per year as it has done for decades. It persists in reflecting the magnificent Pamir peaks and glaciers, and the inhabitants of the Bartang Gorge continue their daily struggle for a livelihood, now beneficiaries of much improved communications and an up-to-date early warning system. I have had no further word about Jack the partridge, but he did teach all of us a lesson – remain calm in any real or apparent crisis!

All would be well if this idyllic scenario were an accurate prediction. Unfortunately, agitation by the news media with their thirst for drama, if not melodrama, and those who abet them, cannot let go of a sellable story. I address this issue in the next chapter.

Fig. 134.   Imja Lake from the air in 2007. The
newly regarded major threat to outbreak of the
lake is the course of drainage through the end
moraine via the series of small lakes (Photograph
kindly supplied by Sharad P. Joshi, ICIMOD,
Kathmandu).

CHAPTER 16

Threatened Disasters in the Pamir and Himalaya

I am very concerned about the melodramatic reactions to potential mountain hazards that seem to have proliferated since the UN declaration of 2002 as the International Year of Mountains – not that I am wishing to suggest any connection between the two, merely the coincidence of timing. After the ebb of alarmism about Himalayan deforestation and the proposed downstream flooding of Gangetic India and Bangladesh, the next major scare with which I was directly involved was the Lake Sarez investigation, detailed in the previous chapter. While we thought that this gross exaggeration had been laid to rest, it appears that the claims about pending catastrophes have persistent lives of their own and that the presumed dire threat stemming from Lake Sarez has been resurrected. Coincidental with this was widespread alarm relating to the assumed precipitous disappearance of all Himalayan glaciers. Initially, it was limited to the perceived immediate danger of catastrophic outbreak of glacial lakes, widely reported in the news media as "glacial lake outburst floods," or GLOFs. Nevertheless, it quickly enlarged to include the much broader issue of the effects of climate warming.

This chapter is devoted to the twin topics of exaggerated reporting about the dangers posed by Lake Sarez and Himalayan glaciers because the alarms follow a similar pattern to the earlier reporting on Himalayan deforestation (Chapters 7 and 10). As previously discussed, the original melodrama is best characterized by the 1979 World Bank report, that by AD 2000 all accessible forest cover in Nepal would be eliminated. The catastrophe assumed to accompany such deforestation, referred to as the theory of Himalayan environmental degradation, today is lost in the mists of time – and that is a rather short time. Yet the end of the millennium came and went and there may have been more forest remaining in Nepal in 2000 than existed in 1979. Nor was there any significant retraction by the agencies and individuals who first reported the alarms, except for a single instance (FAO/CIFOR 2005).[107]

The current crisis-generating spree, once again is largely based more on sentiment than established fact. It is a further reminder of Michael Thompson's provocative "What would you like the facts to be?" The following account is presented, therefore, because it

J. D. Ives, *Sustainable Mountain Development*, https://doi.org/10.1007/978-3-030-96029-2_16

illustrates how my personal mountain journey, the core of this book, appears to have no end in sight.

Lake Sarez: Pamir Aftermath

It ensued that our Pamirian partridge's lesson of remaining calm (Chapter 15) did not penetrate parts of the news media, nor even a section of the supposedly peer-reviewed scientific literature.

Following the April 2003 false alarm and completion of the Swiss/UN early warning systems, there arose renewed interest in the possible danger posed by Lake Sarez. Several international conferences were held in Nurek and Dushanbe, Tajikistan, and annual surveillance of the lake was undertaken by the Swiss engineering firm that installed the early warning systems. The most recent conference was held in Nurek in 2009 when several proposals were debated. High-level delegates from several of the Central Asian republics and Russia urged that the lake level be lowered artificially by 50 m to 80 m, an amount believed to ensure absolute safety. It was proposed that difficulties posed by cutting directly through the unconsolidated landslide material could be avoided by tunnelling the bedrock on the valley side. This would also allow construction of a hydro-electric power station; the energy produced could then be sold to the neighbouring countries and so greatly reduce costs, assuming purchasers could be found.

Given the altitude and remoteness of the lake and the high cost of constructing power lines over great distances through mountainous terrain, this proposal would require the infusion of huge sums of international funding (estimated at about one-half billion US dollars), whether or not the threat was imminent.

A recommendation for a follow-up conference, to be held in 2011 to celebrate the hundred-year anniversary of the lake's original creation, was approved unanimously. Nevertheless, although in less dramatic tone than in some earlier news items, the prospect was raised of an imminent collapse of biblical proportions with millions of lives at risk.

Dr. Jörg Hanisch contacted me with his reaction to this publication. We submitted a joint letter of protest to the editor. It was rejected.

It is widely understood that many natural cataclysmic events are exceedingly difficult, if not impossible, to predict. Thus, avalanches, glacial lake outburst floods, landslides, and giant rockfalls defy precise evaluation. The situation is rendered much more difficult when such natural phenomena are triggered by earthquakes that make prediction even more uncertain. A real problem with such instances is how to balance a decision to go ahead with preventative action on humanitarian grounds against the usually very high costs involved. An overriding issue, however, is the tendency of the news media and even self-serving scientists, administrators, and politicians to seize any opportunity to create public alarm. Unease, panic, even unwise costly preventative measures provoked by public or governmental pressure may be more disruptive than the potential event itself.

Glacial Lake Outburst Floods Revisited

Following the early studies of the UNU team, and especially those of June Hammond and Teiji Watanabi (Chapter 10) on the Dig Tsho 4 August 1985 disaster and Imja Lake in the Khumbu Himal, there was a hiatus in official interest. Two events changed this situation. The first was an alarm spread by the Sherpas of the Rolwaling Himal claiming that a large and rapidly expanding glacial lake was hanging over their villages and appeared to be on the point of overtopping its end moraine dam. It was no exaggeration.

It was widely reported in Nepal that a supra-glacial lake, Tsho Rolpa, was forming on the lower part of the Trakarding Glacier. It had developed over the same time period, but it appeared much more unstable than Imja Lake. The initial alarm produced an immediate response. Much of the valley below was evacuated for several months, a hydro-electric power station much farther down-valley was temporarily closed; even scheduled air flights were suspended for a short period. The longer-term response was for the lake level to be artificially lowered by three metres with financial support from the government of the Netherlands. International consultants were hired, including Dr John Reynolds, who had had extensive experience with similar problems in the Andes, and scientific investigations were undertaken.

Two electronic early warning systems were installed and regular monitoring was set up. This effort collapsed, however, during the Maoist disturbance that left Nepal in the chaos of civil war between 1996 and 2006.

Nevertheless, as a result of the publicity, the entrepreneurial news media began to sniff out a marketable story.

One of the most disturbing examples of news media opportunism is a 2002 report entitled "Meltdown in the Himalaya." In it he quoted John Reynolds as having predicted that "the 21st century could see hundreds of millions dead and tens of billions of dollars in damage" due to the outbreak of glacial lakes worldwide, but principally in the Himalaya and the Andes. Reynolds, however, insists that he was totally misquoted. There were several other claims by the news media that such outburst floods could extend for hundreds of kilometres, cross the borders of Nepal and Bhutan, and cause extensive damage to large Indian cities on the Ganges.

The danger of glacial lake outbursts is real (Ives 1986 2004; Ives et al. 2010), but misquotation and gross exaggeration are totally inappropriate, if not unethical. Unnecessary responses by the national and international authorities and by innocent people living downstream of the glaciers in question can be extremely disruptive. This is discussed further below.

The second event that revived the earlier concern about the glacial lake outburst hazard was the decision by the World Bank, together with Germany and Japan as the major donors, to proceed with construction of a cascade of huge hydro-electric plants along the Arun River that lies immediately east of the Khumbu. The site chosen for the initial dam construction was designated Arun III. It is located in the mid-section of the Arun Gorge below the international frontier with

Tibet (China). More than 90% of the Arun watershed lies in Tibet; a preliminary survey identified several potentially dangerous glacial lakes – as the UNU mountain hazards mapping team had done so a decade earlier. The largest lake to be identified, the Lower Barun Glacial Lake, is located on the Nepal side of the border. Its volume was estimated to be 28 000 000 m³ and a rough calculation indicated that, if its end moraine dam were to collapse, the ensuing flood surge would impact the construction site within about an hour.

In conjunction with the rapid build-up of post-1992 (Tsho Rolpa) glacial lake hazard awareness, the World Bank organized a meeting of experts in Paris in April 1995 to review the situation of Arun III. The group consisted of senior representatives of the donor countries, Germany and Japan, the World Bank, the engineering companies involved in the project design, and Nepal itself (I was gratified to find that our former UNU colleague Pradeep Mool was representing Nepal). The engineering consultant firms were represented by Drs Wolfgang Grabs, Jörg Hanisch, and John Reynolds. Two independent, non-aligned experts were invited. One was a long-time Swiss colleague, glaciologist and engineer, Dr Hans Röthlisberger; the other, to my surprise, although hardly an expert, was me.

At the opening session, it appeared that Hans and I were the only experts present who had strongly negative thoughts about a hydro-electric scheme that would cost many times the GNP of Nepal and that would have a huge environmental impact. Over a private dinner, Hans reflected that our negative reaction was based on our concern for the environment, although that topic was strictly outside the terms of reference of the consultation. He remarked that the World Bank was fortunate to have recruited two non-aligned scientists because it was apparent from the large amount of literature sent to us earlier that the danger posed by the prospect of glacial lake outbursts was minimal and should not prevent a decision to begin construction if that was to be the only issue.

Just before leaving home to fly to Paris, I had received from Teiji Watanabe a manuscript describing his latest research on Imja Lake. It indicated that there was the possibility for a serious outburst, although more detailed research was needed. During the final morning's discussion in the World Bank's Paris offices, I explained the availability of Teiji's manuscript. I was ruled out of order on the grounds that Imja Lake was located in a different watershed to the proposed site of Arun III. Although this ruling was eminently reasonable under the restrictive terms of the meeting, during the ensuing coffee break, the German government representative asked if I would let her have a copy of Teiji's manuscript. With Teiji's subsequent permission, I forwarded a copy to her after my return home from Paris.

The concluding decision was that the Arun III project should proceed; the recommendation was unanimous. Soon after, I learned through contacts in Switzerland that the German government, apparently reluctant to remain involved on environmental grounds, used the threat of Imja Lake to withdraw its support for the project. The reported justification to withdraw was that if Imja Lake should discharge, even though it was in a different watershed, there would be such a high level of public reaction that it would induce widespread opposition to the entire Arun Cascade proposal. Later, it became evident that the German government had been looking for an excuse to withdraw on environmental and economic grounds, but these were more difficult to sustain internationally. The German withdrawal caused the project to collapse.

An additional explanation is that, just at the time of World Bank decision-making on Arun III, James Wolfensohn was elected president; one of the first things he did was to cancel the project, I believe, on economic grounds. The aid money from Germany and Japan, however, was by no means lost to Nepal. Germany built the Middle Marsyangdi and Japan the Kali Gandaki power stations as alternatives.

Himalayan Hazards and Climate Change

The issue of climate warming had only entered the 1992 Rio Earth Summit deliberations in rather general terms, and it was not a primary factor in the United Nations decision to declare 2002 as the International Year of Mountains. Nevertheless, climate warming is now an all-embracing issue and has obvious relevance to our early UNU mountain hazards mapping work in the Khumbu. The inappropriate manner of popularizing the danger posed by rapidly expanding glacial lakes has been illustrated in Chapter 10. Widespread linkage to climate warming did not occur to any great extent before the close of the twentieth century.

Pradeep Mool and his colleagues, working with ICIMOD, were already using satellite imagery to produce inventories of glaciers and glacial lakes in Nepal and Bhutan by the year 2001 (Mool et al. 2001a 2001b) and several university researchers and institutions independently undertook relevant fieldwork, especially in the Khumbu. I was invited by ICIMOD to spend eight weeks between early December 2009 and March 2010, to work with Pradeep and several of his colleagues to write a report for the UN International Strategy for Disaster Reduction (ISDR): *Formation of Glacial Lakes in the Hindu Kush-Himalaya and GLOF Risk Assessment* (Ives et al. 2010). This was followed by a second publication for the World Bank: *Glacial Lakes and Glacial Lake Outburst Floods in Nepal* (Mool et al. 2011). These publications contain strong cautions about the tendency to exaggerate.

In the summer of 2009 before leaving for Kathmandu, I had been working with a group of colleagues to produce a paper entitled "Global warming and its effects on the Himalayan glaciers of Nepal" at the personal request of the editor of an online journal. The manuscript was completed and submitted shortly before I left for Kathmandu in November 2009. My co-authors were well aware of the exaggerated claims that were being spread by the news media (and not only by the news media) about the impacts of global warming on the Himalaya. We made it a central focus of the paper. As an introduction, we inserted a number of the most outrageous claims we could extract from the popular press and other sources. These included the quotation quoted earlier in this chapter together with the following:

> *Himalayan glaciers could vanish within 40 years: 500 million people in countries like India could be at risk of drought and starvation.* (*The Times*, 21 July 2003)

> *Glaciers in the Himalaya are receding faster than in any other part of the world and, if the present rate continues, the likelihood of them disappearing by the year 2035 and perhaps sooner is very high if the Earth keeps warming at the current rate.* (Intergovernmental Panel on Climate Change, Cruz et al. 2007: 493)

There were also political contradictions such as when Shri Jairam Ramesh, the Indian Union Minister for Environment and Forests, gave the address of welcome at a scientific conference in India:

> *… the retreat of Himalayan glaciers is not due to Climate Change … has no scientific evidence and these scenarios are painted by the West.* (*North Indian Times*, 8 September 2009)

The minister claimed that the order of magnitude of retreat by Himalayan glaciers is "a couple of cm to a couple of metres every year … and some are actually advancing." The overall pronouncement is as fallacious as the opposite extreme of the statements that it was intended to discredit.

The "doomsday" year of 2035 eventually was attributed to Professor Syed Hasnain. It had inadvertently slipped into the 2007 report of the Intergovernmental Panel on Climate Change (IPCC) and was widely distributed by the opponents of climate warming immediately before the international conference held in Copenhagen in December 2009. It caused an acrimonious explosion and was one of the reasons for little progress during the conference. Yet, as was the case of quotations attributed to John Reynolds, Syed Hasnain flatly denied ever having made such a prediction.[108]

The discussion had led from the hazards of Himalayan glacial lake outbursts, claiming that millions of lives will be lost, to the subsequent speculation that all Himalayan glaciers would disappear by 2035 (or at least, in the near future). It was also predicted, as a consequence of glacier disappearance, that the major rivers of the region, the Ganges, Indus, Brahmaputra, Yangtze, and others would be reduced to seasonal streams. The next step in the melodrama would be the consequent death of millions due to drought and the collapse of agriculture (for extreme contrast in interpretation, see Alford 2011 and Alford et al. 2011).

Our co-authored manuscript was submitted for publication in October 2009. It contained careful empirical research by Don Alford and Richard Armstrong which, amongst other findings, led to the hypothesis that the total volume of glacier ice in Nepal, if melted instantly, would add only about four to six per cent of the volume of that year's flow of the Ganges. Furthermore, it included Alton Byers's replication of the Fritz Müller/Erwin Schneider map and photographs from the 1950s. One of the photo-pairs shows the Khumbu Glacier; its snout had not retreated visibly between the 1950s and 2008 although appreciable thinning had occurred. A report published by the UN Environment Programme claimed that the Khumbu Glacier had retreated by five kilometres. Our co-authored paper also included Teiji Watanabe's 2009 research on the Imja Glacier from which he concluded that the danger of a catastrophic flood had been exaggerated (Watanabe et al. 2009).

The submitted paper did indicate that new, potentially dangerous, lakes were forming and glaciers were thinning and retreating; some smaller glaciers at lower altitudes had totally disappeared. There is no intention to imply here that climate warming is not reducing glaciers throughout wide areas of the Himalaya. It is! Nevertheless, the gross exaggerations, even falsifications, that we contested should have been self-evident.

The first group of anonymous reviewers approached by the editor of the online journal professed to be too busy to respond (under the circumstances, this appeared remarkable). The total comments and questions of the second group exceeded the length of the paper – in exasperation we withdrew it. I could not avoid the suspicion that there may have been an undertaking not to publish it close to the timing of the Copenhagen conference.

Some of the most recent and insistent representation of the likelihood of imminent large-scale death and destruction is contained in a number of professionally produced videos and by extensive use of the Internet. Results from the original field survey and research stemming from UNU's Khumbu mountain hazards mapping project of the 1980s are rarely encountered. Yet Watanabe

and colleagues had been able to determine that Imja Lake, while its surface area had extended rapidly, had fallen in level by 37 m since 1960. This contrasts with a mere three-metre artificial lowering of Tsho Rolpa in the early 1990s to contain the danger of an outburst.

Imja Lake has been in existence for more than a half century yet no outburst has occurred, despite its continued enlargement. Earthquakes have long been recognized as one of the major processes that could destabilize the end moraine dams in this highly seismic region. Thus when the 6.9 Richter scale earthquake of 18 September 2011 struck Sikkim, eastern Nepal, and adjacent regions, there was concern that glacial lake outbursts would occur. Certainly the earthquake set off landslides and rockfalls, and caused a large amount of damage and loss of life, as far away as Kathmandu, well to the west of Imja Lake, but no precipitous drainage of a glacial lake.

Over the last several years videos have been used to exaggerate the potential large-scale disaster of glacial lake drainage; this has included the U.S. Public Broadcasting System, and most recently, UNDP. The videos are exquisitely filmed and edited and their narrators are accomplished professionals, although not glaciologists. I will use the UNDP video as an example. Produced for UNDP by Arrowhead Films, it is well worth viewing for the beauty of the footage of both the high mountain landscape and the local people. The narrator's first sentence claims that the Himalaya contain 40% of the world's fresh water. This claim is at least an order of magnitude in excess of reality although it is difficult to refute with specific data. This, in large part, is because such data do not exist. The commentary contains no definition of what is meant by "40 per cent of the world's fresh water." If the ice sheets of Antarctica and Greenland are included, as is conventional, and if the volume contained in the North American Great Lakes, northern Canada, Lake Baikal, and the lakes of East Africa is added, also a regular convention, then the claim of the video's first sentence is several orders of magnitude in error. When the first sentence contains such an obvious falsehood, the informed viewer must wonder how much can be accepted from the remainder of the video.

Manipulation by camera is as old as photography itself. I will provide only a single example, chosen from a universally distributed video, because it centres on Imja Lake, widely promoted as the most dangerous glacial lake in the Himalaya. Within the first minute of the video, the camera pans across an unnamed town that is being overwhelmed by a torrent of flood water. Houses are collapsing, vehicles are being washed away, and crowds of pedestrians are desperately struggling to escape. The scene then moves to a picturesque view of Imja Glacier, its lake, and the surrounding high mountains; then back again to people being washed away in an altogether different and unidentified raging flood. The narrator's voice is tense, dramatic, but the three scenes are not causally connected, nor are the instances of flooding identified. It may be assumed that the thousands, if not millions, of anticipated viewers are expected to make the obvious causal connection; but it is false, and we are left with the disturbing thought that there is no relationship between the graphically depicted towns being flooded and any glacial lake outburst.

The message of this chapter is twofold. First, society has been subjected to outrageous and alarmist exaggerations, even to the point of deliberate falsification, apparently by organizations and individuals, some of whom are generally perceived as responsible authorities. Second, the current alarms with which we have all been bombarded since at least 2002 have paralleled the earlier conventional wisdom of the 1970s and 1980s claiming that, by 2000, there would be no accessible forests left in Nepal (World Bank 1979, Asian Development Bank 1982), and that Gangetic India and Bangladesh would be under water after thousands of landslides, induced by 'ignorant' mountain farmers, had stripped both vegetation and soil off the Himalaya. Regardless, Hamilton's early provocative statement, "it floods in Bangladesh when it rains in Bangladesh" warrants repetition. This common sense reaction by Larry has since been substantiated by an extension of the UNU mountain project, largely funded by Switzerland (Hofer and Messerli 2006) and applies with equal force to the current claims that melting of all the Himalayan glaciers would significantly reduce the flow of the Ganges; it is the monsoon rainfall that supplies the overwhelming volume of water to the Ganges and Brahmaputra rivers, not glacier melt.

When we began our Himalayan quest in 1978 we never expected that we would become involved in two intellectual-political controversies: first, deforestation and environmental and socio-economic collapse; second, loss of all the Himalayan glaciers and consequent flooding, followed by a regional drought and the death of millions. Following the outcry in Copenhagen in 2009, the claims that Himalayan glaciers will disappear in the near future has at least helped to prompt an urgent request for much needed new research, systematic data collection, and rational thinking. Research is now rapidly expanding to ensure more accurate assessment of the interface between climate change and glacier and snow cover response. Nevertheless, it hasn't arrested the onslaught of exaggeration and confusion, now by widespread distribution of professional videos. Yet we are left with the very serious question – how to predict the likelihood of catastrophic natural events, in this case, glacier lake outburst floods, when there is no simple nor reliable method. To underplay the hazard would be equally irresponsible. This calls for the re-introduction of the word "dilemma" that initially emerged from the Mohonk Process. Regardless, the time has long passed when the local mountain people should be incorporated into the research activities and their extensive environmental knowledge introduced as a vital component in the search of practical solutions.

Aftermath on Imja Lake

Since the foregoing critical assessment was written, Alton Byers of The Mountain Institute, in association with ICIMOD, led an international expedition to inspect the problematic Imja Lake. The group included Peruvian engineers who had had extensive experience in the secure management of dangerous glacial lakes in the Andes.

One of the results of the inspection has been a partial reassessment of the danger of a lake outburst. The expedition was accompanied by Suzanne Goldenberg. *The Guardian* published her report on 10 October 2011 as "Glacier Lakes: Growing danger zones in the Himalayas." To quote:

> *The extent of recent changes to Imja Lake has taken glacial experts by surprise, including Teiji Watanabe …. He said that he did not expect such rapid changes to the moraine which is holding back the lake … "We need action, hopefully within five years. I feel our time is shorter than what I thought before. Ten years might be too late."*

By subsequent email exchange, Teiji indicated that he had not seen the report. However, he did point out to me that changes to the lake outlet through the moraine were occurring faster than he had previously observed, confirming in general the remarks attributed to him.

Obviously, such situations in nature do change unexpectedly. Imja Lake may be less stable than we have recently assumed. As was indicated in the 2010 ICIMOD publication (Ives et al. 2010), a glacial lake outburst could occur tomorrow. Imja Lake could take ten years or more to burst; or it could drain slowly and safely; Dig Tsho did burst in 1985, resulting in several deaths and destruction of a small hydro-power station nearing completion. Nevertheless, the present situation requires planning and constant observation. It does not justify excessive alarmism or false reporting, and that is the primary message of this chapter.[109]

Photographs on the facing page
Figures 135 and 136 are included here for final emphasis despite a degree of repetition (see page 171). This is because Imja Lake has become a *cause célèbre* for the news media and many others who prophesize an almost immediate outburst with dramatic loss of life and property. What is required instead of melodrama is regular inspection while more detailed geophysical work is continued in close cooperation with the local inhabitants.

Fig 135.   The upper Imja Glacier from the flanks of 'Island Peak' as seen in 1956. (Photograph from the collection of Fritz Müller)

Fig. 136.   In recent years Dr Alton C. Byers has made replication of old Himalayan and Andean glacier photographs almost an obsession, attracting other photographers to do the same. This photograph, taken above Imja Lake, Khumbu Himal, looks toward Ama Dablam, lost in cloud to left of figure (Alton Byers). It is included here as a tribute to Dr Byers, in characteristic pose, and to be shared with Professor Teiji Watanabe.

Fig. 137. A 1980 scene on the Tibetan Plateau showing the remnants of an ancient abandoned fortified settlement. Was the abandonment caused by over-grazing, drought induced by climate warming in the distant past, or both?

CHAPTER 17

The Evolution of the Mountain Cause

In this book, I have outlined the evolution of the mountain cause from the perspective of a small group of highly committed scholars as it emerged from purely academic research collaboration into international environmental politics. I chose the mid-1960s as the point of departure for this narrative because, until then, there had been no international cohesion of mountain research, and environmental and development politics were only beginning to gather strength. So mountains were perceived as having little significance in the big issues of the day, such as North–South relations, population growth, poverty, desertification, destruction of equatorial rainforest, and pollution of the oceans. Mountains did not even merit a footnote in the 1972 Stockholm UN Conference on the Human Environment.

The Stockholm conference, however, was the first international event of its kind. It set the stage for the impressive upsurge of concern for environmental issues of the next 40 years, yet mountain regions and the peoples who inhabited them were not part of the early agenda. In retrospect, it appears that this oversight or lack of concern may have been due either to their remoteness, seemingly inviolate ruggedness, and lack of significant connection to mainstream world affairs, or to the assumption that their problems would be adequately addressed piecemeal under the broad banner of concerns such as deforestation and the depletion of other major ecosystems of the world.

In expounding this story of growing mountain awareness from 1964 onward, I have used a personal approach because it has allowed me to describe my involvement with the persons and personalities and leading scholars who played a vital role in that story. Likewise, our meetings and relationships with people who might appropriately be described as world leaders were often on a strikingly personal basis. The personal approach also enabled me to describe many contacts with poor mountain people who were virtually anonymous or frequently dismissed as ignorant peasants in need of being shown the way. I sincerely hope the book demonstrates that the mountain peoples, in their local settings, are finally recognized as essential parts of any agenda aimed at attaining a comfortably inhabitable Planet Earth.

J. D. Ives, *Sustainable Mountain Development*, https://doi.org/10.1007/978-3-030-96029-2_17

The narrative approach also enlivens the mountain journey with human experiences that in reality and memory frequently transcend the mundane juggling of international agency politics, the arduous challenges of day-to-day mountain research, and the publication and debate of scientific findings. The "journey" I have related shows how a relatively small group of determined scholars were empowered to push their highly controversial research into the political arena and so contribute to the success of major objectives such as the United Nations designation of 2002 as the International Year of Mountains. As it evolved, this required not only collaboration with many institutions, agencies, and experts, but also a number of fortuitous happenings and chance meetings with influential people that had repercussions beyond anything that could have been expected at the outset of the mountain trek.

Major goals emerged during the course of the journey; the political objectives could hardly have been envisaged in the early years simply because the environmental movement itself unfolded only gradually. It remains to be seen whether the enthusiasm and sense of achievement of 2002 will be reattained during the far more complicated and disruptive disputes of 2012 under the rubric of Rio+20. This account is written in a period of severe worldwide economic difficulty, international strain, and military threat. Nevertheless, many of the early effects of climate change can be clearly identified in the world's high mountain regions. Thus, the relevance of sustainable mountain development is much more widely accepted. The major problem that will face the post-Rio+20 years will be to find a way to balance measures to ease current global economic and political instability with attempts to resolve the longer-range environmental issues, including an all-embracing response to climate change.

The accounts in this book also illuminate the tendency, even determination, of some of the world's most powerful agencies as well as influential journalists, editors, and science writers to exaggerate perceived catastrophe. Exaggeration contains the potential to induce confusion, to cause serious loss of time for proper identification of real cause and effect, and so delay, or even thwart, appropriate responses. Unfortunately the "identify-a-catastrophe" game appears to have become embedded in present-day mass psychology, from the presumed dangers of travelling by air to the threat of pandemics and the immediacy of the next giant earthquake or tsunami. This frame of mind is self-defeating, except for its unfortunate by-product of enhancing the news media or inflating the budget of some agency or NGO. The mountain journey, as described in these pages, has attempted to demonstrate the inadequacy of such false alarms.

The primary lesson to be learned about mountain awareness must await the outcome of the results of the June 2012 Rio+20 UN Conference. Whether or not the well-being of mountain peoples and the environmental sustainability of their homelands will be advanced in any significant way will not be known with certainty for several years. Nevertheless, it is safe to say that, from the perspective of this book, the major stepping stones of the journey so far – the IGU Mountain Commission 1968, UNESCO MAB-6 1973, UNU Mountain Project 1978, Mohonk Conference 1986, Rio Earth Summit 1992, UN headquarters review 1997, International Year of Mountains 2002 – have helped greatly to promote worldwide awareness of the importance of mountains to the future security of society at large. This has been further advanced by the accelerated formation of institutions, NGOs, and the expansion of scientific research since 2002. Yet it must be realized that there is always a strong temptation for world leaders to put off hard decisions on seemingly long-term goals in deference to immediate economic and security concerns.

From an entirely personal viewpoint, the mountain journey described here has been exceptionally rewarding. Yet it is apparent that it has no end. It is comforting to know, therefore, that many of the students of the original group of mountain academic researchers, their students and colleagues, and the many mountain peoples who have since emerged onto the world scene, have now firmly assumed the role of guardians of the mountains.

Fig. 138.   I encountered this small girl in her island village on Lake Titicaca, Bolivia. She is introduced here as part of our hope for the future leaders of the mountains, their indigenous inhabitants, and to emphasize the necessity of treating them as equal collaborators and leaders in the commitment for sustainable mountain development.

Epilogue

Martin F. Price

The preceding chapters were completed in early 2012. In June of that year, the Rio+20 conference took place. This epilogue considers, first, the attention given to mountains at that conference and subsequently in global processes focussing on sustainable development. The next section discusses a number of international organisations that were either introduced earlier in the book or have emerged following activities that have been described, with an emphasis on the decade since the first edition of this book was published. This epilogue concludes with comments on current and long-term issues that need to be considered in the context of sustainable mountain development.

Mountains on the Global Stage

Mountains at Rio+20

The outcome document of the Rio+20 conference, 'The Future We Want[1]', contains 283 non-binding paragraphs, of which three refer specifically to mountains.

Mountains

210. *We recognize that the benefits derived from mountain regions are essential for sustainable development. Mountain ecosystems play a crucial role in providing water resources to a large portion of the world's population; fragile mountain ecosystems are particularly vulnerable to the adverse impacts of climate change, deforestation and forest degradation, land use change, land degradation and natural disasters; and mountain glaciers around the world are retreating and getting thinner, with increasing impacts on the environment and human well-being.*

211. *We further recognize that mountains are often home to communities, including indigenous peoples and local communities, who have developed sustainable uses of mountain resources. These communities are, however, often marginalized, and we therefore stress that continued effort will be required to address poverty, food security and nutrition, social exclusion*

---

1 Available at https://sustainabledevelopment.un.org/content/documents/733FutureWeWant.pdf

and environmental degradation in these areas. We invite States to strengthen cooperative action with effective involvement and sharing of experience of all relevant stakeholders, by strengthening existing arrangements, agreements and centres of excellence for sustainable mountain development, as well as exploring new arrangements and agreements, as appropriate.

212. *We call for greater efforts towards the conservation of mountain ecosystems, including their biodiversity. We encourage States to adopt a long-term vision and holistic approaches, including through incorporating mountain-specific policies into national sustainable development strategies, which could include, inter alia, poverty reduction plans and programmes for mountain areas, particularly in developing countries. In this regard, we call for international support for sustainable mountain development in developing countries.*

One might consider that the inclusion of only three paragraphs indicates that the signatories of the document—Heads of State and Government and high-level representatives—did not recognize the importance of mountains to the same extent as at UNCED in 1992, whose final document—'Agenda 21'—devoted one of its 40 chapters to sustainable mountain development, as described in Chapter 13 of this book. Nevertheless, it is notable that other global issues—such as climate change, biodiversity and desertification, each with a

UN Convention signed at UNCED—receive similar attention in 'The Future We Want'. Consequently, it may be argued that the three paragraphs on mountains did show continued global attention to sustainable mountain development and reflect progress in the mountain cause.

To provide context for this statement, it is necessary to go back to 1993, when FAO was designated the lead agency for implementing the mountain chapter of 'Agenda 21'. As described in Chapter 13 of this book, from 1993 onwards, FAO organised a series of meetings that led, among other things, to the establishment of the Mountain Forum in 1995 and the International Year of Mountains 2002. 2002 was also the year of the Rio+10 conference, the World Summit on Sustainable Development (WSSD). For the mountain cause, this meeting was important because it included the founding of the Mountain Partnership: a 'voluntary alliance of partners dedicated to improving the livelihoods of mountain people and protecting mountain environments around the world'.[2] The key players in its establishment were the Governments of Switzerland and Italy, FAO, and UNEP; effectively, it provided a formal structure to replace the informal one that had existed since 1993. The Mountain Partnership is coordinated by FAO and brings together a wide range of mountain stakeholders, including governments as well as international, intergovernmental and scientific organisations and NGOs and private sector companies active in many fields. From the end of 2012 to late 2021, its membership has more than doubled, from

---

2 https://www.fao.org/mountain-partnership/about/en/.

200 members from 73 countries to 437 members from 96 countries.

During the preparations for Rio+20, the Mountain Partnership Secretariat played a key role in bringing together many of its members—notably, the founding governments and organisations, ICIMOD, the Centre for Development and Environment (CDE) at the University of Bern, and the Mountain Research Initiative—to draft the text on mountains. Other input came from the 'Major Groups' of the UN Commission on Sustainable Development and negotiators at the Preparatory Committee meetings for the WSSD. Thus, the three paragraphs may be considered the product of a broad consultative process which recognised that the challenges identified in 1992 were still largely current, and further exacerbated by climate change. This process was complemented by the preparation and publication of a series of ten regional reports on progress towards sustainable mountain development from 1992 to 2012, also coordinated by the Mountain Partnership.[3]

Mountains and the Sustainable Development Goals

Three years after the Rio+20 conference, the UN convened a high-level plenary meeting of its General Assembly, in New York, to adopt the post-2015 development agenda, titled 'Transforming Our World: The 2030 Agenda for Sustainable Development'.[4] This includes 17 Sustainable Development Goals (SDGs)

and 169 targets. As for the previous conference, the Mountain Partnership and its members conducted a very active campaign over the preceding two years, advocating for the inclusion of text specific to mountains. The outcome is that mountains are mentioned, first in paragraph 33 of the final document, which refers to the conservation and sustainable use of 'oceans and seas, freshwater resources, as well as forests, mountains and drylands', and also in three targets. One of these refers specifically to mountains. Under SDG 15, which considers the sustainable use of terrestrial ecosystems, target 15.4 is:

By 2030, ensure the conservation of mountain ecosystems, including their biodiversity, in order to enhance their capacity to provide benefits that are essential for sustainable development.

Mountains are also mentioned in target 15.1:

By 2020, ensure the conservation, restoration and sustainable use of terrestrial and inland freshwater ecosystems and their services, in particular forests, wetlands, mountains and drylands, in line with obligations under international agreements;

and also under SDG 6, on freshwater: target 6.6 is:

By 2020, protect and restore water-related ecosystems, including mountains, forests, wetlands, rivers, aquifers and lakes.

---

3 The regional reports considered the mountains of Africa; the Alps; the Andes; Central Asia; Central, Eastern, and Southeastern Europe; the Hindu Kush-Himalaya; Meso America; the Middle East and North Africa; North America; Southeast Asia and the Pacific. They are available at https://www.fao.org/mountain-partnership/publications/policybriefs/en/.
4 Available at https://sdgs.un.org/2030agenda.

Again, while four specific mentions of mountains may appear few in number, the 2030 agenda mentions other geographic entities, such as drylands and deserts, a similar number of times. The three targets show continued global attention to the mountain cause, particularly on the topics of biodiversity conservation, ecosystem services, and water resources.

Since 2015, considerable work has been done on the relevance of not only these SDGs and their targets, but many others, for sustainable mountain development. An issue brief published by CDE and the Mountain Research Initiative[5] considers this in relation to five mountain countries—Ecuador, Kyrgyzstan, Nepal, Switzerland and Uganda—and concludes that, despite considerable differences between the mountain regions of these countries, there are also common development priorities. However, this document and a related working paper considering Bangladesh, Chile, Ecuador and Nepal[6] both conclude the lack of disaggregated data to sufficiently distinguish the situation in these areas from that at the national scale—at which the SDGs are typically assessed and reported—remains a key challenge for action in mountain areas. This challenge was recognised three decades ago in Chapter 13 of 'Agenda 21'; although our knowledge and understanding of mountain environments and people has increased significantly, there is much work yet to be done.

Keeping Mountains on the Agenda

While the availability of disaggregated data is a continuing challenge with regard to many of the 'facts' of sustainable mountain development, data for human populations at the global scale are reasonably robust. Consequently, the Mountain Partnership Secretariat has been able to prepare three reports that consider the distribution of mountain people around the world and the underlying factors of vulnerability that many of them face. The latest report[7] concludes that, in 2017, the global mountain population was nearly 1.1 billion, 15 percent of the world's population—and that this represented an increase of 89 million people since 2012, almost all in developing countries. Crucially, 346 million of these people—half of those living in the mountains of developing countries—were vulnerable to food insecurity: their daily availability of calories and protein is below the minimum threshold for a healthy life. These are fundamental issues for sustainable mountain development.

Keeping mountains on the global agenda requires showing their relevance to global processes. Consequently, the Mountain Partnership Secretariat, together with other members, has published reports that show the relevance of mountains in relation to the issues addressed by: the global conventions on climate change, biodiversity, and desertification; International Years, such as the

5 Wymann von Dach S, Bracher C, Peralvo M, Perez K, Adler C (2018) Leaving no one in mountains behind: Localizing the SDGs for resilience of mountain people and ecosystems. Bern: Centre for Development and Environment and Mountain Research Initiative, with Bern Open Publishing.

6 Bracher C, Wymann von Dach S, Adler C (2018). Challenges and opportunities in assessing sustainable mountain development using the UN Sustainable Development Goals. Bern: Mountain Research Initiative and Centre for Development and Environment.

7 Romeo R, Grita F, Parisi F, Russo L (2020) Vulnerability of mountain peoples to food insecurity: updated data and analysis of drivers. Rome: Food and Agriculture Organisation of the United Nations.

International Year of Family Farming (2014); the Sendai Framework for Disaster Risk Reduction (2015–2030); and the 2030 Agenda.[8] In addition, the Secretariat and FAO prepare reports on sustainable mountain development for the UN Secretary General, which provide valuable summaries of the activities of governments and organisations around the world. Over the past decade, these reports have been presented in 2013, 2016 and 2019 to the UN General Assembly, which has then adopted a Resolution on this topic.[9] All of these reports show that mountains are increasingly important in key global contexts and that the number of governments and organisations active in the mountain cause has continued to grow, and the resolutions underline that sustainable mountain development remains on the global agenda. This was further substantiated when, in December 2021, the UN General Assembly adopted a resolution, sponsored by 94 governments, declaring that 2022 would be the International Year of Sustainable Mountain Development.[10]

Mountain Organisations

Among the members of the Mountain Partnership are many international organisations, some of which have been mentioned in previous chapters. This section focuses on five of these organisations, recognising that there are many others which have been and are active,

to a greater or lesser extent, on issues related to sustainable mountain development. In addition, there are many regional and national organisations active in this field.

In the Hindu Kush Himalaya (HKH), ICIMOD, introduced in Chapter 3, continues as a unique intergovernmental organisation that is supported by the governments of eight countries[11] in a region where cooperation is often not strong, if even existent, and there is a history of conflict which, in some cases, continues. Nevertheless, ICIMOD staff from all eight countries work together with the aim of developing and sharing 'research, information and innovations to empower people' in the mountains of these countries. ICIMOD provides 'a regional platform where experts, planners, policymakers and practitioners can exchange ideas and perspectives towards the achievement of sustainable mountain development'.[12] Its three strategic impact areas are reduced poverty, enhanced resilience by reducing physical and social vulnerabilities, and enhanced ecosystem services. In addition to many activities on these topics over the past decade, a particularly notable accomplishment has been the publication of the HKH Assessment in 2019, following five years of collaborative work by more than 350 scientists from the HKH and beyond.[13] It sets a global benchmark for the integrated assessment of the available knowledge for a large mountain region and also presents scenarios

8 These are available at https://www.fao.org/mountain-partnership/publications/mountain-partnership-key-publications/en/.

9 These are available at https://www.fao.org/mountain-partnership/publications/un-documents/en/.

10 https://www.fao.org/mountain-partnership/publications/publication-detail/en/c/1460891/.

11 Afghanistan, Bangladesh, Bhutan, China, India, Myanmar, Nepal and Pakistan.

12 ICIMOD (2017) Strategy and results framework. Kathmandu: International Centre for Integrated Mountain Development.

13 Wester P, Mishra A, Mukherji A, Shrestha AB (eds) (2019) The Hindu Kush Himalaya assessment—Mountains, climate change, sustainability and people. Cham: Springer Nature Switzerland AG.

for the future of the region, in the context of the SDGs. One measure of the global recognition of the importance of this region is that the funding for the assessment came from the governments of not only the eight countries of the HKH, but also Australia, Austria, Sweden, Switzerland and the UK. Within the region, the assessment was used as the basis for consultative processes in the eight countries, resulting in a 'Call for Action' with specific actions towards a shared vision that the future of the region should be 'one in which its societies and its people—children, women and men—are prosperous, healthy, peaceful and resilient in a healthy environment'.[14]

In the past decade, two international organisations have been increasingly active and are introduced here because of their focus on collaboration and research: the Mountain Research Initiative (MRI) and the Global Mountain Safeguard Research (GLOMOS) programme of the UNU Institute for Environment and Human Security (UNU-EHS) and Eurac Research. The MRI was established in 2001 and has been primarily supported by a number of Swiss institutions ever since. The MRI 'envisions a world in which research to identify and understand drivers and processes of global change in mountains is promoted and linked across disciplines and mountain regions worldwide'.[15] Effectively, it provides a means to facilitate comparative worldwide research on mountain issues, as called for at the MAB-6 Lillehammer meeting in 1973,

described in Chapter 2 of this book. Importantly, since its establishment nearly three decades later, the MRI has been allied with the successive international programmes on research on, and monitoring of, global change: first, the International Geosphere-Biosphere Programme (IGBP), the International Human Dimensions Programme (IHDP), and the Global Terrestrial Observation System (GTOS); and subsequently Future Earth and the Group on Earth Observations (GEO). An early activity of the MRI was to publish a book—Global Change and Mountain Regions: An Overview of Current Knowledge[16]—which, in many ways, updated 'Mountains of the World: A Global Priority', published in 1997 (see Chapter 13).

As mentioned above, the MRI collaborated with CDE on the assessment of sustainable mountain development in the context of the SDGs from 2013 to 2018. In addition, over the past decade, the MRI has played a vital role in facilitating globally coordinated research, usually working with other active networks and organisations, resulting in key global publications on biodiversity, high-elevation climates, glacier recession and meltwater, and ecosystem services in mountain areas.[17] These topics are all connected, and considered in an MRI-led initiative with GEO, initially launched in 2016 as the GEO Global Network for Observations and information in Mountain Environments (GEO-GNOME) and renamed in 2021 as GEO Mountains. Its

14 ICIMOD (2020) The HKH Call for Action to sustain mountain environments and improve livelihoods in the Hindu Kush Himalaya. Kathmandu: ICIMOD.

15 https://www.mountainresearchinitiative.org/who-we-are.

16 Huber UM, Bugmann HKM, Reasoner MA (eds) (2005) Global change and mountain regions: An overview of current knowledge. Dordrecht: Springer.

17 The resulting publications may be found at https://www.mountainresearchinitiative.org/activities/mri-publications.

aim is to 'bring together research institutions and monitoring networks to enhance the discoverability, accessibility and usability of a wide range of relevant data and information pertaining to environmental and socio-economic systems—both in situ and remotely sensed—across global mountain regions'.[18] It should therefore lead to further progress in addressing the perennial need for mountain-specific data at all spatial scales. As a GEO initiative, GEO Mountains is designed to support the 2030 agenda, Sendai Framework and the UN Framework Convention on Climate Change. In the latter context, MRI, other Swiss-based institutions[19] and ICIMOD have been leading the preparation of a chapter on mountains for the sixth assessment of the Intergovernmental Panel on Climate Change (IPCC), the first IPCC report to include such a chapter since the second assessment in 1994. MRI activities include not only research and knowledge exchange, but also capacity-building; MRI has a working group on education for sustainable development[20] and was also closely involved in four global conferences of mountain scientists.[21] Such activities are essential not only for providing opportunities to survey the state of the art, and gaps, in mountain research worldwide, but also for ensuring the long-term continuity of communities of active scientists and informed practitioners

and policymakers for sustainable mountain development.

GLOMOS was established in 2019 in North Tyrol, Italy, and represents a renewed attention to mountain issues within the UNU two decades after the conclusion of activities within its 'Highland-Lowland Interactive Systems' (later, 'Mountain Geoecology and Sustainable Development') project, described from Chapter 6 onwards, and a subsequent project in the Pamir-Alai.[22] GLOMOS aims to be 'an interface between the international mountain research community and the UN system'. Its goal is to 'contribute to the development of resilient mountain communities towards natural and man-made hazards and disaster risks, to protect the wealth of biological and cultural diversity, and to support adaptive solutions and sustainable transformation processes within these highly sensitive social-ecological systems, first and foremost in the Global South'.[23] As well as activities in Ecuador and South Africa,[24] a major current project is the preparation of an edited book, to be published in 2022, that will provide a state-of-the-art overview of research, identify challenges and enhance understanding of the diversity of mountain contexts in regions around the world, and present the recognition of mountain issues within international frameworks. This will be the first

---

18 https://www.geomountains.org/rationale-and-objectives/rationale.

19 University of Zurich, Helvetas.

20 This published a special issue (Vol 40, 4) of MRD in 2020.

21 In Perth, Scotland in 2005, 2010 and 2015, organised by the Centre for Mountain Studies, Perth College, University of the Highlands and Islands; and in 2019, organised by the University of Innsbruck. The former two conferences both resulted in special issues of MRD (Vol. 32, S1; Vol. 36, 4), as well as other publications.

22 Pachova NI, Renaud FG, Hirsch D, Anarbaev M, Mamatov T, Ergashev M, Olimov I (2012) Towards sustainable land management in the Pamir-Alai Mountains. Bonn: United Nations University Institute for Environment and Human Security.

23 https://ehs.unu.edu/about/departments/glomos#overview.

24 Szarzynski J, Delves JL, Fontanella Pisa P, Membretti A, Robles SP, Pedoth L, Schneiderbauer S (2020) The Global Mountain Safeguard Research (GLOMOS) Programme: Linking academia and the United Nations system for transformative resilience in mountain regions. Mountain Research and Development 40(4): P4–P7.

overview of global mountain research and key issues since 2005.

Both the MRI and GLOMOS are partners in a new initiative on mountain issues within UNESCO's MAB Programme, the World Network of Mountain Biosphere Reserves, launched in 2021 following a decade during which there were no coordinated mountain activities within this programme[25]. In addition, the MRI and GLOMOS—as well as other organisations including FAO, ICIMOD and CDE—are members of the International Mountain Society, which continues to publish Mountain Research and Development (MRD). Since it was founded in 1981 by Jack Ives, who was its first editor, as described in previous chapters, the scope of the journal has continued to expand[26]. While there are other international journals with a mountain focus, MRD is pre-eminent in its field. It is unique in publishing not only peer-reviewed research, but also development approaches and experiences from mountains around the world as well as evidence-based agendas for research and policy; and actively fostering science-practice-policy dialogue and capacity-building for scientists in developing countries.

The range of organisations active in 'getting the facts right', disseminating them and increasing the capacity of people around the world to work towards sustainable mountain development has increased significantly over the past decade, as shown by the growing membership of the Mountain Partnership.

This epilogue has only specifically referred to six of these organisations. IMS and ICIMOD have been in existence for some decades; MRI and the Mountain Partnership have been active for two decades; and GLOMOS and the new MAB initiative show the renewed interest of UNU and UNESCO in mountain issues. There is significant collaboration between these organisations, and many of their activities are undertaken jointly with other organisations, of many different types, active at various scales. The majority of all the active organisations are members of the Mountain Partnership, whose membership includes the governments of 60 countries. The extent to which they are committed to sustainable mountain development, both in their own countries and more widely, varies. Nevertheless, to resume a theme introduced earlier in this book, Switzerland has continued to play a particularly important role in the mountain cause and is a major contributor to the activities of the Mountain Partnership, ICIMOD, MRI and IMS; the latter two are both located at the University of Bern.

Key Issues for Sustainable Mountain Development

The concept of sustainable mountain development was first introduced at the international scale in 1992, in Chapter 13 of 'Agenda 21'. This specified two 'programme areas' that needed to be addressed towards achieving this goal: 'generating and strengthening

25 Price MF, Schaaf T, Cárdenas Tomažič MR, Fontanella Pisa P, Köck G (2022) The World Network of Mountain Biosphere Reserves. In Schneiderbauer S, Szarzynski J, Fontanella Pisa P, Shroder JF (eds) Safeguarding mountains: A global challenge. Amsterdam: Elsevier (in press).

26 https://www.mrd-journal.org/about/.

knowledge about the ecology and sustainable development of mountain ecosystems' and 'promoting integrated watershed development and alternative livelihood opportunities'.[27] Ten years later, in 2002, a series of background papers were prepared for the Bishkek Global Mountain Summit, the final global event of the International Year of Mountains.[28] These identified 10 sets of key issues for sustainable mountain development. It is beyond the scope of this epilogue to comment on subsequent progress, though the reports on sustainable mountain development prepared for the UN General Assembly provide useful summaries, and it is to be hoped that the forthcoming book being coordinated by GLOMOS will provide an up-to-date assessment. More recently, as part of the work done in 2015-2018 by CDE and MRI on sustainable mountain development in the context of the SDG targets, a number were identified as particularly relevant. These relate to climate change and its impacts; the resilience of mountain people and ecosystems; the conservation and sustainable use of these ecosystems; the eradication of poverty; sustainable tourism; and health and education. All of these were recognised as important in 2002, and all but climate change in Chapter 13 of 'Agenda 21' ten years earlier. A key conclusion of this recent work comes back to many of the themes explored in this book: 'Given the high sensitivity of mountains to climate change, the many disaster risks, and the diversity of priorities, entry points must be found for policies and interventions that simultaneously *address critical development issues and strengthen the resilience of mountain people and ecosystems'.*[29]

This epilogue has been written two years into the coronavirus disease (COVID-19) pandemic, which has very clearly shown how important it is that mountain people and ecosystems are resilient, or can be supported to move in this direction. In addition to its direct health and mortality effects, COVID-19 has had many other impacts on mountain people. In the HKH, these have included increases in psychological stress, gender-based violence, and gender, social and economic inequalities; and the disruption or collapse of food systems and supply chains. The last of these is one of the impacts that shows the extent to which mountain societies and economies are entwined with global economies; others are the loss of remittance incomes and the cessation of tourism.[30] All of these impacts are likely to have been experienced in mountain communities in all developing countries and—apart from remittances and in countries where the tourism industry is more dependent on domestic than international travel—worldwide. As economies recover and travel becomes more possible, some of these impacts are starting to become less severe, though the pandemic will have many long-term effects.

27 United Nations (1992) Managing Fragile Ecosystems: Sustainable Mountain Development. Chapter 13 in Agenda 21, Rio de Janeiro: United Nations Conference on Environment and Development.

28 These were published as Price MF, Jansky L, Iatsenia AA (eds) (2004) Key issues for mountain areas. Tokyo: UNU Press.

29 Wymann von Dach S, Bracher C, Peralvo M, Perez K, Adler C (2018) Leaving no one in mountains behind: Localizing the SDGs for resilience of mountain people and ecosystems. Bern: Centre for Development and Environment and Mountain Research Initiative, with Bern Open Publishing. Italics in original.

30 ICIMOD (2020) COVID-19 impact and policy responses in the Hindu Kush Himalaya. Kathmandu: International Centre for Integrated Mountain Development.

The other pre-eminent global context is climate change, which is having—and will continue to have—increasing impacts on mountain ecosystems, and the people who live in them, around the world. The frequency and intensity of 'natural disasters'—very often influenced by human activity, as described in previous chapters—is increasing. Such trends are of existential significance not only for mountain people but for billions living downstream, and show the importance of recognising the connections between all the different global challenges and crises that we face. Unfortunately, the outcomes and commitments of COP26, held in Glasgow in November 2021, seem unlikely to have been substantive enough to decrease emissions of greenhouse gases to limit global warming to 1.5°C. The world is becoming a more uncertain place. Ever more than before, there is a need for governments, international organisations, non-governmental organisations and private businesses to cooperate in the interests of sustainable mountain development—and it is to be hoped that the International Year of Sustainable Mountain Development, 2022, will catalyse and strengthen such cooperation, with valuable long-term outcomes. A secure future for the mountains will benefit the entire world.

APPENDIX I

List of Acronyms

CAS	Chinese Academy of Sciences (Academia Sinica)
CIA	Central Intelligence Agency (USA)
CIDA	Canadian International Development Agency
CNRS	Centre National de la Recherche Scientifique (France)
FAO	Food and Agriculture Organization (UN)
GTZ	German Agency for Technical Cooperation
IBP	International Biological Programme
ICIMOD	International Centre for Integrated Mountain Development
ICSU	International Council of Scientific Unions
IDRC	International Development Research Centre (Canada)
IGC	International Geographical Congress
IGU	International Geographical Union
IIASA	International Institute for Applied Systems Analysis
IMS	International Mountain Society
INGO	International non-governmental organization
INSTAAR	Institute of Arctic and Alpine Research (subsequently Arctic, Antarctic and Alpine Research)
IPCC	Intergovernmental Panel on Climate Change (UN)
IUCN	International Union for the Conservation of Nature and Natural Resources
KGB	Committee for State Security (former USSR)
MAB	Man and the Biosphere Programme (UNESCO)
MIT	Massachusetts Institute of Technology
MRD	*Mountain Research and Development* (journal)
NCAR	National Center for Atmospheric Research (USA)
NGO	non-governmental organization
NOAA	National Oceanic and Atmospheric Administration (USA)

© The Author(s), under exclusive license to Springer Nature Switzerland AG 2022
J. D. Ives, *Sustainable Mountain Development*, https://doi.org/10.1007/978-3-030-96029-2

NRC	National Research Council (USA)
NSF	National Science Foundation (USA)
RNAC	Royal Nepal Airlines Corporation
SATA	Swiss Association for Technical Assistance
SDC	Swiss Agency for Development and Cooperation
STOL	Short take-off and landing (aircraft)
UN	United Nations
UNCED	United Nations Conference on Environment and Development
UNDP	United Nations Development Programme
UNEP	United Nations Environment Programme
UNESCO	United Nations Educational, Scientific and Cultural Organization
UNU	United Nations University
USAID	United States Agency for International Development
WHO	World Health Organization (UN)

(names of several of the agencies changed throughout the period covered)

APPENDIX II

Mohonk Mountain Conference Resolution

RESOLUTION 1

The Mohonk Mountain Conference recognizes the special role of the Himalaya as a unique part of the world cultural heritage and wishes to draw attention to the critical importance of its spiritual contribution to the well-being of the world community.

RESOLUTION 2

The Mohonk Mountain Conference reaffirms that a serious situation has been developing in the Himalayan Region for several decades. This relates to the progressive environmental deterioration and the pronounced decline in the standard of living of many of the peoples affected, particularly the mountain peoples. One aspect is the rapid increase in total population in relation to the available agricultural and forest land. Taking this into consideration, it is resolved that an international conference be convened as soon as possible to further examine the issues and to recommend an urgent course of action.

It is also resolved that a small working group be formed to develop an action research design and to lay the groundwork for the proposed conference. The working group should be formed by the International Mountain Society in co-operation with the United Nations University, the East-West Center, the International Union for the Conversation of Nature and Natural Resources, the International Centre for Integrated Mountain Development (ICIMOD), and other organizations.

RESOLUTION 3

Realizing that nature recognizes no international boundaries and that many of the issues and challenges facing development and conservation cannot be dealt with adequately without co-operation between the countries of the Himalayan Region, the Mohonk Mountain Conference strongly urges the governments of the Himalayan Region to take steps to establish international parks in border areas (Parks for Peace) to promote peace, friendship, and co-operation in research and management, for the optimal sustainable use of the natural and human resources, and to improve the quality of life of all the peoples of the region.

J. D. Ives, *Sustainable Mountain Development*, https://doi.org/10.1007/978-3-030-96029-2

RESOLUTION 4

The Mohonk Mountain Conference endorses efforts by the International Centre for Integrated Mountain Development (ICIMOD) to develop a documentation centre and to improve the dissemination of vital information on the region, particularly relating to hydrology and sediment transfer. It is recommended that these efforts be accelerated and that links be established with other appropriate institutions.

RESOLUTION 5

The Mohonk Mountain Conference welcomes the recent initiative of the World Resources Institute, the World Bank, and the United Nations Development Programme in establishing a tropical forest action plan which should facilitate efforts to deal with the comprehensive land-use aspects of the problem facing the Himalayan Region, and calls upon donors to provide the appropriate support.

CONCLUSION 1

The Mohonk Mountain Conference concludes that the Himalayan Region is best characterized as one of great complexity, not amenable to generalized development policies and panaceas. In view of this, the extensive current research on many aspects of the region should be accelerated and emphasis should be placed on specific and inter-related issues. These include, amongst others:

- the status and role of women in rural areas
- organized and spontaneous migration of peoples
- population growth patterns
- indigenous production strategies
- land ownership and taxation patterns
- the development and effectiveness of indigenous self-help movements
- co-ordinated research on selected trans-mountain watersheds

CONCLUSION 2

Taking into account the accumulating knowledge and expertise of its worldwide membership and publication system, the International Mountain Society resolves to make itself available, in conjunction with other organizations, to undertake research and to advise upon project proposals developed for application to the Himalayan Region.

Mohonk Mountain House, 11 April 1986
Under the chairmanship of Maurice Strong

References

Alford, D. (2011). Hydrology and glaciers in the Upper Indus Basin. World Bank Technical Report, forthcoming.

Alford, D., Armstrong, R., and Racoviteanu, A. (2011). Glacier retreat in the Nepal Himalaya: The role of glaciers in stream flow from the Nepal Himalaya. World Bank Technical Report, forthcoming.

Arrowhead Films (2011). *Himalayan Meltdown.* Video at http://arrowheadfilms.com/channel/himalayan-meltdown/ [pwd: meltdown]

Asian Development Bank (1982). *Nepal Agricultural Sector Strategy Study,* 2 vols. Asian Development Bank: Kathmandu.

Brower, B. (1996). Geography and history in the Solukhumbu landscape. *Mountain Research and Development,* 16 (3): 249–256.

Buchroithner, M.F., Jentsch, G., and Wanivenhaus, B. (1982). Monitoring of recent geological events in the Khumbu area (Himalaya, Nepal) by digital processing of Landsat MSS data. *Rock Mechanics,* 15: 181–197.

Byers, A.C. (1986). A geomorphic study of man-induced soil erosion in the Sagarmatha (Mount Everest) National Park, Khumbu, Nepal. *Mountain Research and Development,* 7 (3): 83–87.

Byers, A.C. (1987a). Landscape change and man-accelerated soil loss: The case of the Sagarmatha (Mount Everest) National Park, Khumbu, Nepal. *Mountain Research and Development,* 7 (3): 209–216.

Byers, A.C. (1987b). *Geoecological study of landscape change and man-accelerated soil loss: The case of Sagarmatha (Mount Everest) National Park, Khumbu, Nepal.* Unpublished doctoral dissertation, Department of Geography, University of Colorado.

Byers, A.C. (2005). Contemporary human impacts on alpine landscapes in the Sagarmatha (Mount Everest) National Park, Khumbu, Nepal. *Annals of the Association of American Geographers,* 95 (1):112–140.

Caine, N. and Mool, P.K. (1981). Channel geometry and flow estimates for two small mountain streams in the Middle Hills, Nepal. *Mountain Research and Development,* 1 (3–4): 231–243.

Caine, N. and Mool, P.K. (1982). Landslides in the Kolpu Khola drainage, Middle Mountains, Nepal. *Mountain Research and Development,* 2 (2): 157–173.

Chinese Academy of Sciences (1981). *Geological and Ecological Studies of Qinghai-Xizang Plateau.* Volume 1: *Geology, Geological History and Origin of Qinghai-Xizang Plateau,* pp.1–974. Volume 2: *Environment and Ecology of Qinghai-Xizang Plateau,* pp. 975–2138. Chinese Academy of Sciences, Science Press: Beijing, and Godon and Breach: New York.

Cruz, R.V. et al. (2007). *Asia. Climate Change: Impacts, Adaptation and Vulnerability.* Contribution of Working Group II of the Fourth Assessment Report of the IPCC. Cambridge University Press, p. 493.

Davidson, F.P. (1983). *MACRO: A Clear Vision of How Science and Technology Will Shape Our Future.* William Morrow and Co., New York.

Eckholm, E. (1975). The deterioration of mountain

environments. *Science,* 189: 764–770.

Eckholm, E. (1976). *Losing Ground.* Worldwatch Institute. W.W. Norton and Co.: New York.

Forsyth, T.J. (1996). Science, myth, and knowledge: Testing Himalayan environmental degradation in Northern Thailand. *Geoforum,* 27 (3):375–392.

Forsyth, T.J. (1998). Mountain myths revisited: integrating natural and social environmental science. *Mountain Research and Development,* 18 (2): 107–117.

Fürer-Haimendorf, C. von (1975). *Himalayan Traders: Life in Highland Nepal.* John Murray: London.

Hammond, J.E. (1988). Glacial lakes in the Khumbu region, Nepal: An assessment of the hazards. Unpublished Master's thesis, Department of Geography, University of Colorado: Boulder.

Hinrichsen, D., Lucas, P.H.C., Coburn, B., and Upreti, B.N. (1983). Saving Sagarmatha. *Ambio,* 11 (5): 274–281.

Hofer, T. and Messerli, B. (2006). *Floods in Bangladesh: History, dynamics and rethinking the role of the Himalayas.* United Nations University Press: Tokyo.

Imhof, E. (1974). *Die Grossen Kalten Berg von Szetschuan.* Orell Füssli Verlag: Zurich.

Ives, J.D. (1970). Himalayan highway. *Canadian Geographical Journal,* 80 (1): 26–31.

Ives, J.D. (1986). Glacial lake outburst floods and risk engineering in the Himalaya. Occasional Paper no. 5. ICIMOD: Kathmandu.

Ives, J.D. (1987). The theory of Himalayan environmental degradation: Its validity and application challenged by recent research. *Mountain Research and Development,* 7 (3): 189–199.

Ives, J.D. (2004). *Himalayan Perceptions: Environmental change and the well-being of mountain peoples.* Routledge: London and New York. 2nd edition, 2006. Himalayan Association for the Advancement of Science: Kathmandu.

Ives, J.D. (2010). *The Land Beyond: A Memoir.* University of Alaska Press: Fairbanks, Alaska.

Ives, J.D. (2014). Research expeditions to Baffin Island, Canadian Arctic: 1961 to 1967.

Ives, J.D. and Ives, P. (eds.) (1987). The Himalayan-Ganges Problem: Proceedings of the Mohonk Mountain Conference. *Mountain Research and Development,* 7 (3): 181–344.

Ives, J.D. and Messerli, B. (1989). *The Himalayan Dilemma: Reconciling development and conservation.* Routledge: London and New York.

Ives, J.D., Shrestha, R.B. and Mool, P.K. (2010). *Formation of Glacial Lakes in the Hindu Hush-Himalayas and GLOF Risk Assessment.* International Centre for Integrated Mountain Development, Kathmandu.

Johnson, K., Olsen, E.A., and Manadhar, S. (1982). Experimental knowledge and response to natural hazards in mountainous Nepal. *Mountain Research and Development,* 2 (2): 175–188.

Kienholz, H., Hafner, H., Schneider, G. and Tamrakar, R. (1983). Mountain hazards mapping in Nepal's Middle Mountains with maps of land use and geomorphic damages (Kathmandu-Kakani area). *Mountain Research and Development,* 3 (3): 195–220.

Kienholz, H., Schneider, G., Bichsel, M., Grunder, M. and Mool, P. (1984). Mountain hazards mapping project, Nepal: Base map, and map of mountain hazards and slope stability, Kathmandu-Kakani area. *Mountain Research and Development,* 4 (3): 247–266.

Lamsal, D. Sawagaki, T. and Watanabe, T. (2011). Digital terrain modelling using Corona and ALOS PRISM Data to investigate the distal part of the Imja Glacier, Khumbu Himal, Nepal. *Journal of Mountain Science,* 8 (3): in press.

Mahat, T.B.S., Griffin, D.M. and Shepherd, K.R. (1986 a). Human impact on some forests of the Middle Hills of Nepal: 1, forestry in the context of the traditional resources of the state. *Mountain Research and Development,* 6 (3): 223–232.

Mahat, T.B.S., Griffin, D.M. and Shepherd, K.R. (1986b). Human impact on some forests of the Middle Hills of Nepal: 2, some major human impacts before 1950 on the forests of Sindhu Palchok and Kabhre Palanchok. *Mountain Research and Development,* 6 (4): 325–334.

Mahat, T.B.S., Griffin, D.M. and Shepherd, K.R. (1987a). Human impact on some forests of the Middle Hills of Nepal: 3, forests in the subsistence economy of Sindhu Palchok and Kabhre Palanchok. *Mountain Research and Development,* 7 (1): 53–70.

Mahat, T.B.S., Griffin, D.M. and Shepherd, K.R. (1987b). Human impact on some forests of the Middle Hills of Nepal: 4, a detailed study in southwest Sindhu Palchok and Kabhre Palanchok. *Mountain Research and Development*, 7 (2): 111–134.

Manandhar, S. (Gurung) (2005). *Beyond the Myth of Eco-Crisis: Local Responses to Pressure on Land in Nepal. A Study of Kakani in the Middle Hills*. Mandala Publications: Kathmandu.

Mool, P.K., Bajracharya, S.R. and Joshi, S.P. (2001a). *Inventory of Glaciers, Glacial Lakes and Glacial Lake Outburst Floods. Monitoring and Early Warning Systems in the Hindu Kush-Himalayan Region – Nepal*. International Centre for Integrated Mountain Development: Kathmandu.

Mool. P.K.. et al. (2001b). *Inventory of Glaciers, Glacial Lakes and Glacial Lake Outburst Floods. Monitoring and Early Warning Systems in the Hindu Kush-Himalayan Region – Bhutan*. International Centre for Integrated Mountain Development: Kathmandu.

Mool, P. K. et al. (2011). *Glacial Lakes and Glacial Lake Outburst Floods in Nepal*. International Centre for Integrated Mountain Development: Kathmandu.

Moseley, R.K. (2006). Historical landscape change in northwest Yunnan, China. *Mountain Research and Development*, 26 (3): 214–219.

Moser, P. and Moser,W. (1986). Reflections of the MAB-6 Obergurgl Project and Tourism in the Alpine Environment. *Mountain Research and Development*, 6 (2): 101–118.

Müller, F. (1958). Eight months of glacier and soil research in the Everest region. *The Mountain World 1958/59*. Allen and Unwin: London, pp. 191–208.

Pearce, F. (2002). Meltdown. *The New Scientist*, 2 November, 2002, pp. 44–48.

Pye-Smith, C. (1990). *Travels in Nepal: The sequestered kingdom*. Car Books: Bristol, UK.

Rerkasem, B. (ed.) (1996). *Montane Mainland Southeast Asia in Transition*. Chiang Mai University:Thailand.

Schuster, R.L. and Alford, D. (2000). *Usoi Landslide Dam and Lake Sarez*. United Nations, ISDR Prevention Series No. 1, United Nations, New York and Geneva.

Sicroff, S. (1998). *Approaching the Jade Dragon: Tourism in Lijiang County, Yunnan Province, China*. Unpublished master's thesis, University of California, Davis.

Sutton, S.B. (1974). *The China's Border Provinces: The Turbulent Career of Joseph Rock, Botanist and Explorer*. Hastings House, New York.

Swope, L.H. (1995). *Factors influencing rates of deforestation in Lijiang County, Yunnan Province, China*. Unpublished master's thesis, University of California, Davis.

Swope, L.H., Swain, M.B., Yang, F. and Ives, J.D. (1997). Uncommon property rights in southwest China: Trees and tourists. In R.B. Johnston (ed.) *Life and Death Matters: Human Rights and the Environment at the End of the Millennium*. Altamira Press, Sage Publications: Walnut Creek, California.

Thompson, M. (1995). Policy-making in the face of uncertainty: The Himalayas as unknowns. In G.P. Chapman and M. Thompson (eds.), *Water and the Quest for Sustainable Development in the Ganges Valley*. Mansell: London, pp. 25–38.

Thompson, M and Warburton, M. (1985). Uncertainty on a Himalayan scale. *Mountain Research and Development*, 5 (2): 115–135.

Thompson, M., Warburton, M. and Hatley, T. (1986). *Uncertainty on a Himalayan Scale*. Ethnographia: London.

UNESCO – MAB. (1977). Regional meeting on integrated ecological research and training needs in the southern Asian mountain systems, particularly in the Hindu Kush-Himalayas. MAB report series No. 34, Paris.

Vuichard, D. (1986). Geological and petrographic investigations for the Mountain Hazards Project, Khumbu Himal, Nepal. *Mountain Research and Development*, 6 (1): 41–51.

Vuichard, D. and Zimmermann, M. (1986). The Langmoche flash-flood, Khumbu Himal, Nepal. *Mountain Research and Development*, 6 (1): 90–94.

Vuichard, D. and Zimmermann, M. (1987). The catastrophic drainage of a moraine-dammed lake, Khumbu Himal, Nepal: Cause and consequences. *Mountain Research and Development*, 7 (2): 91–110.

Watanabe, T. (1994). Soil erosion on Yak grazing steps in the Langtang Himal, Nepal. *Mountain Research and Development*, 14(2): 171-179.

Watanabe, T., Ives, J.D. and Hammond, J.E. (1994). Rapid growth of a glacial lake in Khumbu Himal, Nepal: Prospects for a catastrophic flood. *Mountain Research and Development*, 14 (4): 329–340.

Watanabe, T. Kameyama, S. and Sato, T. (1995). Imja Glacier dead-ice melt rates and changes in a supra-glacial lake, 1989–1994, Khumbu Himal, Nepal: danger of lake drainage. *Mountain Research and Development*, 15 (4): 293–300.

Watanabe, T., Lamsal, D. and Ives, J.D. (2009). Evaluating the growth characteristics of a glacial lake and its degree of danger: Imja Glacier, Khumbu Himal, Nepal. *Norsk Geografisk Tiddskrift*, 62: 255-267.

Zimmermann, M., Bichsel, M., and Kienholz, H. (1986). Mountain hazards mapping in the Khumbu Himal, Nepal, with prototype map, scale 1:50,000. *Mountain Research and Development*, 6 (1): 29–40.

Endnotes

1   Quotation from *The Bangladesh Observer*, Dhaka, 2 June 1990 under the heading "Deforestation in the Himalaya Aggravating Floods."

2   Quotation from World Resources Institute, "Tropical forests: A call to action." Report of an international task force convened by the World Resources Institute, the World Bank, and the United Nations Development Programme, Washington DC: WRI, 1985.

3   The Cold War, the major political and military standoff between the Soviet Union and its satellites and the North Atlantic Alliance, which developed after World War II, seeped its antagonisms even to the extent of distorting international scientific cooperation. The so-called non-aligned countries, often referred to as the Third World, were also caught up in the deadly competition; many of them, for various political and/or economic reasons, would vote at the United Nations in the interests of the Soviet bloc. This had descended into international scientific affairs. In London in 1964, Academician Gerasimov threw the International Geographical Union into disarray by precipitously renouncing his position as its senior vice-president on a political expedient related to the East German delegation (Soviet bloc) not receiving their UK visas in time to enable them to attend (Chapter 2).

4   Roger G. Barry, Lawrence Hamilton, Dipak Gyawali, Dinesh Dhakal, Mel Marcus, Bruno Messerli, Sun Honglie, David Griffin, Ruedi Höger, Kevin O'Connor, David Pitt, Mischa Plam, Michael Thompson, and others.

5   Maurice Strong, prior to our first contact with him, had been a dominant figure in international environmental politics. One of his more significant roles was to work with Gro Harlem Brundtland, then prime minister of Norway, to produce the first major clarion call to threats facing our environment (*Our Common Future* was produced through the United Nations in 1975). He had previously served as secretary-general of the United Nations Conference on the Human Environment (Stockholm 1972) leading to establishment of the United Nations Environment Programme, headquartered in Nairobi, of which he was the founding director-general.

6   Prominent amongst them were Yuri Badenkov, Jayanta Banyopadhyay, Tim Forsyth, Robert Rhoades, Fausto Maldonado, Jane Pratt, Fausto Sarmiento, Inger Marie Bjønness, Peter Stone, and a growing group of former graduate students of the first generation: Don Alford, Barbara Brower, Alton Byers, Carol Harden, Hans Hurni, Hans Kienholz, Martin Price, Teiji Watanabe, and Robert Zomer.

7   Geomorphology is the term given to a sub-discipline of geography that involves the scientific study of landscape, its classification, and how it evolved. As it also involves analysis of slope processes, it leads into applied studies of landslides, avalanches, floods, and soil erosion. In many, especially northern, countries a significant part of geomorphology takes in the impacts of the Ice Ages on landscape evolution, as well as present-day

glaciology. In short, it is an important part of physical geography and, as such overlaps with climatology, geology, soil mechanics, hydrology, physics, and many other aspects of earth, or natural, science.

8   The 1968 visit to Darjeeling had such an impact on my future approach to mountains and mountain people that a full account is included in Chapter 4.

9   Nel Caine and Kathleen Blackie (Salzberg) already had been offered positions with me in Ottawa and were diverted to Boulder, Colorado.

10   The journal subsequently changed its name to *Arctic, Antarctic, and Alpine Research.*

11   Other members of the "executive committee" of the commission were professors Bruno Messerli (Switzerland), Masatoshi Yoshino (Japan), Wilhelm Lauer (West Germany), Mieczslaw Hess (Poland), and Dr Rimma P. Zimina (USSR; wife of Academician Gerasimov). Professor Peter Hollermann (Bonn University) was recruited as secretary.

12   In the 1970s, international political protocol divided the world into "the West" (developed countries), "the East," and "the Third World." Later, delicacy asserted itself, and Third World countries were apportioned between "least developed," and "developing" countries – of course, additional changes have since been assessed as even more politically correct.

13   This led to our introduction to Walter Moser's remarkable Obergurgl Model, the first attempt to model human and physical interaction in mountain environments.

14   MAB-6 (Salzburg), UNESCO's MAB publ. 8, was in print by 5 June 1973, and di Castri realized that he had a willing "slave" for future episodes, of which there were many.

15   The 1970s were the early days of modelling and computer programming. Most academics were barely emerging from punch cards and tapes. The critically important Obergurgl Model, introduced later, was little more than a classroom flowchart – but its importance resulted from its clearly demonstrated objective of multi- and interdisciplinary thinking. Today it could be claimed that the Obergurgl Model led into what is becoming known as transdisciplinary thinking.

16   The United Nations Environment Programme was one of the main achievements of Maurice Strong and the Stockholm Conference. In 1973, it was the latest of the series of UN programmes and has remained very active until the present day. It was the first of the UN agencies to be headquartered in Africa – Nairobi, Kenya.

17   Troll's conference in Mexico (1966) that dealt with the high-altitude geoecology of Central and South America was the first international high mountain scientific gathering. It was supported by UNESCO.

18   The entire MAB Programme was not intended to be a source of funding for large-scale international research, which would be far beyond either UNESCO's interests or resources. The overall objective was to provide project outlines that would serve as catalysts for interested national and international groups to define specific objectives and use the prestige obtained from the UNESCO umbrella the better to compete for in-country research funds. Many of the 14 MAB projects quickly led to the establishment of national committees that were inspired into action. The U.S. national committee, representing several of the individual MAB projects, including Project 6, was based in the Department of State, Washington, DC.

19   Di Castri explained that, while UNESCO could never publicly influence such an appointment (democracy was the order of the day), he would "arrange" for my formal nomination through "proper" channels. To my eventual surprise, I found that I was nominated by Professor O.V. Makeev (USSR) and seconded by Dr Luis Briancon (Bolivia), neither of whom I had previously met.

20   At the time (1970s), it was traditionally required within the UN agencies to give equal standing to the French and English languages. This often caused problems because many of the "other world" participants could manage English, but not French. The point raised here simply underlines one of the obstacles in international relationships that had to be overcome. On more than one occasion, I was referred to as "Jacques Yves"! That problem no longer exists.

21   This meeting was held in Vienna (5–7

December 1973) after the Lillehammer meeting and its concluding report was incorporated into the Lillehammer proceedings as Appendix 9. I was also invited to attend the Vienna meeting.

22   Many of these kinds of working international meetings were taken as opportunities for host countries and cities to display their virtues as generous hosts. I always felt that, while not strictly necessary, the extracurricular activities served to help knit disparate and far-travelled participants and ensure a lot of enthusiasm and hard work.

23   A couple of years later, I was ten minutes late for the opening of one of our IGU mountain commission meetings in Innsbruck chaired by Carl Troll. I was the last to arrive, on the morning train from Vienna. Troll, a stickler for order and promptness, simply asked, "Yack, vat opera vas it this time?" as he indicated an empty chair beside him.

24   I wonder if any other country would have accepted a foreign national as a member of an official delegation to an international conference.

25   Plans were well advanced for the Mendoza meeting when Ricardo telephoned me at home in Boulder to indicate that it would have to be cancelled because of political problems – his house had been fortuitously sprayed with machine-gun fire the previous night. On learning of the tension in Mendoza, Francesco, from Paris, quickly arranged a substitute meeting in Bogotá.

26   This report materialized in a manner totally unanticipated at the time. By 1981, we were able to establish a new quarterly journal (*Mountain Research and Development*). Separate issues were devoted to the four-part report on the Andes with Paul as guest editor and partially funded by UNESCO and UNEP. This was critical, both to launching the journal and producing in high-quality print an English-language version that UNESCO was also able to use as the base for the French and Spanish editions.

27   The group that provided the impetus for GTZ to initiate the Munich meeting were as follows: John Cool of Harvard University, who was soon to become a major player with the Ford Foundation and to take up residence in Kathmandu; Frank P. Davidson, a macro-engineer and chairman, System

Dynamics Steering Committee, MIT; A.D. Moddie, vice-president of the Himalayan Club, New Delhi; Joseph A. Stein, an American architect and president, Joseph Allan Stein Assoc., New Delhi; and B.B. Vohra, Ministry of Agriculture and Irrigation, New Delhi. They all played influential roles in the account that follows.

28   International Workshop on the Development of Mountain Environment: An Interdisciplinary Approach for a Future Strategy. No date (1975). Ed. Klaus Müller-Hohenstein, German Foundation for International Development.

29   The Club of Rome was an influential global think tank. It was founded in 1968 and comprised a large group of world citizens: current and former heads of state, UN bureaucrats, high-level politicians, scientists, and economists. One of its most influential publications was *The Limits to Growth* (1975).

30   Frank was a most original thinker; he described himself as a macro-engineer (Davidson 1983) and ardent supporter of such major engineering works as the Channel Tunnel and a similar approach to ensure linkage between several Japanese islands. He used his independent position to assist with many developments that will be introduced later in the book, although it was not until the turn of the millennium that the Oxford English Dictionary endorsed the derivative term "montology." Following the meeting, we both had a spare evening awaiting our respective flights back to Boston and Denver the following day. This gave me the benefit of a dinner invitation and a long conversation with Frank. A common cause was established that subsequently led to invitations to Boston and a long period of vital advice and assistance, especially during formation of the International Mountain Society and development of the Mohonk conferences (Chapter 12).

31   In the 1960s and 1970s, Sino-Indian rivalry was at an all-time high. Aside from the border war, India had launched a self-interested aid programme of road-building in Nepal. This provided Kathmandu with the first road links to the south and an extensive east-west system. Not to be outdone, the Chinese drove a highway through the

Himalaya to Kathmandu, thereby counteracting any Indian strategic advantage. The Chinese road required a large bridge where it crossed the frontier between Tibet (Xizang Autonomous Region) and Nepal; it was named Friendship Bridge.

32   Mr M'Bow, UNESCO director-general, was personally involved, such was the sense of urgency.

33   Surely an idea assimilated by Professor Walther Manshard and interwoven into the programme of the newly created (1975) United Nations University.

34   In many ways, especially during intensive fieldwork, it was a challenge to mesh Western notions of behaviour, diet, and social propriety with those of contrasting cultures. But it was also a superb opportunity to generate a much higher level of understanding and appreciation. It was also a two-way process.

35   This led to a kind of commercial enterprise once I realized that our neighbours in Boulder, staff members of INSTAAR, and even some of my students, were more than anxious to acquire similar rugs. I first proposed to the refugee centre manager that I could charge twice $64 per rug and still have no problem in selling them. This he refused – principles again getting in the way. I eventually got the better of him by proposing that the surcharge was to assist with the education of the Tibetan children. I ended up selling more than 50 rugs before my link with him faded. But a much more lasting undertaking, also related to educating Tibetan refugee children, quickly followed.

36   At the time, I had received a very warm invitation to spend the year with Professor Gunnar Hoppe in Stockholm. It was a hard decision to make, but once made, it marked the final step in the change of focus in my research activities from Arctic geomorphology to mountains and people.

37   This wayside rest place off the main highway no longer exists.

38   Our sojourn of 1976–77 at Appenberg proved pivotal to several of the main themes of this book. My stay there led to my introduction to Professor Walther Manshard who had been appointed vice-rector of UNU in 1976 and who invited me to accept the position of coordinator of a new university mountain programme; it also was the location in 1990 where the mountain core group met to plan our approach to the Rio Earth Summit and create the *Mountain Agenda* leading to more intense political activism. I liked to compare it to a similar Swiss rural haven of peace and tranquillity, the town hall of Zimmerwald, close to the Messerli family house, where a certain ornithologist by the name of Lenin met in 1917, but not to discuss birds.

39   From the Caucasus research station, we were introduced to the glaciology, landscape processes, such as avalanches and landslides, the local flora and fauna, with the most experienced Russian scholars as guides: Vladimir Kotlyakov (who later succeeded Gerasimov as director of the USSR Institute of Geography), Genady Golubev (who visited Bruno and myself the following year in Switzerland), P.L. Gorchakovsky, R.I. Zlotin, and V.I. Turmanina. All of this in excellent weather and accompanied with outstanding food and colourful peasant dance entertainment in the evenings.

40   Eventually, after I had reached Switzerland I composed a series of letters describing in detail my experience, the plight of the Plams and my vivid recollections from reading the documents. I mailed copies to several U.S. newspapers and wrote a special letter to the president of the United States. I received no acknowledgement from either the White House or the newspapers, although the Plams' fortunes did take a decided turn for the better, as I later relate.

41   One Russian joke describes the Ljubljanka building, headquarters of the KGB, as the tallest building in Moscow because Siberia can be seen even from its basement.

42   On my return to Bern, I was able to confirm Roger Barry's proposal that, because the current director of INSTAAR's Mountain Research Station, David Greenland, would much rather be located close to the main campus, we provide Mischa with highly relevant employment as well as good accommodation for Olga and son.

43   Gerardo Budowski, a former director-general of IUCN, was coordinator of Project 2: Agro-Forestry Systems.

44   Other Chiang Mai workshop participants included Pisit Voraurai, Michel Bruneau, Eric Chapman, Terry Grandstaff, Peter Hoare, Peter Kunstadter, Jack McKinnon, Bruno Messerli, Masatoshi Yoshino (who also had been recruited as a member of our IGU mountain commission), Harald Uhlig, and many Thai colleagues.

45   All the documents were edited and produced under the title *Conservation and Development in Northern Thailand* (Ives, Sabhasri, and Vorauri 1980), UNU Press, Tokyo.

46   Following our retirement and return to Ottawa in 1997, it became apparent that production of a quarterly mountain journal was beginning to exceed our capacity. The transfer of the journal to the Geographical Institute, Bern University, together with a huge infusion of Swiss government funding was something of a personal triumph. With the start of the new millennium, following a 20-year struggle, we were delighted to be succeeded by Hans Hurni and his editorial team.

47   I had intended this as an ideal case for possible future repeat photography so that any further geomorphic activity could be recorded graphically. It became a vital piece of evidence once I realized how determined the local farmers were to ensure complete agricultural recovery.

48   The Chipko Movement (hug the trees) was an Indian Himalayan villagers' movement (initially women) led by Sunderlalji and others, many of whom had been followers and disciples of Mahatma Gandhi. It was a massive protest movement against logging mountain forests, especially resulting from commercial and provincial government corruption. At the height of its strength, it persuaded Prime Minister Indira Gandhi to place a ten-year ban on all Himalayan tree cutting in India.

49   At this point, I must record that, as the project in Nepal developed, Kamal performed yeoman service. Without his assertive approach, many vital things would not have happened or would have been much delayed – as foreign guests, there was no way in which we could have been sufficiently insistent.

50   Dr Schild served as executive director of ICIMOD from 2005 to 2011 and invited me to work with his staff, including Pradeep Mool, to produce a publication on the problem of glacial lake outburst floods.

51   Pradeep's career has been exceptional. He has distinguished himself as a prime mover of Himalayan glacier mapping and analysis of the dangers associated with potential glacier lake outbursts. In this work, I have been associated with him through ICIMOD as recently as 2011.

52   The decision to begin fieldwork on the Kakani ridge actually had very positive rewards. It quickly led us to begin the major challenge to what became known as the theory of Himalayan environmental degradation.

53   I had applied for a visa for the military district of Sikkim from the Indian embassy in Washington DC well before my scheduled departure. After a very long delay and no visa, I received a formal letter explaining the delay and was told I would probably be allowed entry if I showed the letter together with my UN credentials.

54   It was not long since the "outside" world had been startled by President Nixon's visit to China and "ping-pong diplomacy."

55   A very strong Chinese liquor.

56   In 1904, Sir Francis Younghusband led a British military force into Tibet. While Tibet was essentially autonomous, if not independent, at the time, it produced a storm in international relations, including with China.

57   A swift round of telephone calls confirmed that many of our growing group of mountain activists would also meet in Beijing and Lhasa: Bruno Messerli, Corneille Jest, Makato Numata, Harald Uhlig, and Peter Wardle.

58   Roger Barry and Mischa Plam both separated from INSTAAR at the same time, Roger with the World Data Centre for Snow and Ice, together with the staff, transferred to the Cooperative Institute for Research in the Environmental Sciences (CIRES), Mischa to work as a private consultant.

59   Shades of Gilbert and Sullivan.

60   Proceedings of a Symposium on Qinghai-Xizang (Tibet) Plateau, Beijing, 25 May to 1 June 1980. Vol. I: Geology, Geological History and Origin of Qinghai-Xizang Plateau, Vol. II: Geological and

Ecological Studies of the Qinghai-Xizang Plateau, 2138 pp., Science Press, Beijing, and Gordon and Breach Science Publishers, New York, 1981.

61 Because I could not believe that my first, virtually accidental, shot had even been aimed correctly, I asked permission to pose them. Then I exposed the remaining eight frames. The first, accidental, shot proved perfect, the others only run-of-the-mill.

62 Another of many points of surprise showing how Western knowledge may not always be what it seems.

63 Tingri was one of the staging points of the early Mount Everest expeditions of the 1920s from which the expedition members obtained their first distant views of the mountain.

64 Academia Sinica is referred to subsequently as Chinese Academy of Sciences or CAS.

65 These constituted the overview of current environmental and socio-economic problems of the Andes that had been recommended during the MAB-6 regional meeting in La Paz. I was relieved that Paul Baker accepted my invitation to serve as series editor.

66 The first slate of IMS officers were president, JDI; vice-presidents, Corneille Jest and Heinz Löffler; secretary, Roger G. Barry; treasurer, Mischa Plam. The first issue of the journal was published in May 1981: editor, JDI; assistant editor, Pauline Ives; book review editor, Michael Tobias. The editorial advisory board consisted of Paul T. Baker (USA), Roger G. Barry (USA), Gerardo Budowski (Costa Rica), Frank Davidson (USA), Corneille Jest (France), Huang Pingwei (China), Heinz Löffler (Austria), Ricardo Luti (Argentina), Walther Manshard (Germany-UNU), Bruno Messerli (Switzerland), Sanga Sabhasri (Thailand), and Masatoshi Yoshino (Japan).

67 Joseph Rock was a pre-World War II phenomenon closely associated with the National Geographic Society. He took up permanent residence in a Naxi village close to Lijiang and from there travelled extensively. He made rare trips back to the United States that included a dinner at the White House. He was something of a recluse to Western life. He produced a Naxi–English dictionary and made extensive studies of the Naxi way of life. From our point of view, he took a large number of photographs which, although developed in his Naxi village, were of very high quality. Many of them were ultimately archived by the National Geographic Society and became a vital source for subsequent photo-replication that allowed determination of landscape change over a 50-year period. A recent biography penetrates the activities and thoughts of this very odd personality who was eventually thrown out of China at the beginning of Chairman Mao's regime. He was accused of being a spy for the CIA, a fantastic charge (see Sutton 1974 for a biography of Rock).

68 Imhoff, E. (1974). *Die Grossen Kalten Berge von Szetchuan*. Orell Füssli Verlag: Zurich.

69 The "Love-Suicide Meadow" is a fascinating part of Naxi lore dating back to the seventeenth century when the Han emperor decided to bring minority subject nations under stricter control. The military commander in Dayan had been instructed to ensure a closer adherence to Han conventions. This meant that the marriages of Naxi (and other minority) children had to be fixed at birth and the hitherto Naxi concepts of free love and teenage self-selection banned. For a strongly matriarchal society, this produced a remarkable reaction. Young lovers who were about to be separated into a forced marriage responded by going into the mountains, often in groups and, after several days of feasting and love-making, committed suicide in the belief that they would be taken across the mountains to a form of Naxi "Shangri-La" where they would remain eternally young and survive to a life of beauty and ease. Our particular meadow had been one of the favourite "take-off" points for the everlasting life, and the practice had continued into the Communist era.

70 These meetings and field studies included Japan, Ethiopia, Madagascar, the Moroccan Atlas Mountains, the Pyrenees, New Zealand, Australia, Tajikistan, Ecuador, Peru, and Chile. The Swiss team undertook extensive applied research in Ethiopia, Madagascar, and on the Chilean Altiplano. This almost overwhelming series of undertakings certainly left us with a remarkable broadening of

our appreciation of mountain problems that provided solid backing for the eventual push towards Rio and the Earth Summit.

71   Our essential base map for the Kakani area mapping had the unusually valuable scale of 1:10,000 and a ten-metre contour interval. It had been published in 1977 by the German mountain research organization *Arbeitsgemeinschaft für vergleichende Hochgebirgsforschung.*

72   One of the few disadvantages of operating under the auspices of UNU was that the organization was a university and to become involved in actual physical development projects was totally beyond its remit.

73   Actually, this led to a highly relevant field project in Ecuador. With contacts in Quito, I was taken on what was becoming the usual kind of reconnaissance that afforded a traverse across the entire width of the Ecuadorian Andes from the Amazon rainforest, via the high volcanoes, to rain forest on the Pacific coast. In this endeavour, I was accompanied and guided once again by Gerardo Budowski. This reconnaissance was followed by the fieldwork of two graduate students, Carol Harden and Deborah Bossio who, together with Professor Christoph Stadel, undertook separate studies of the range of soil losses under differing altitudinal and farming systems. One of the highly relevant results included a great improvement in knowledge of the sources and rates of reservoir siltation.

74   It proved possible to undertake this essential third part of our original proposal by an extension into Bangladesh, supported by the Swiss Agency for Development and Cooperation (Hofer and Messerli 2006).

75   Playing for time worked sufficiently to give Walther Manshard chance to find a way. He developed an effective alternative: he arranged for UNU to cover the expense of providing complimentary subscriptions for developing world institutions and individuals. IMS was reimbursed for these arrangements to approximately the level of the original UNU publication grant.

76   While the colour proofs of the Khumbu hazard maps were being edited in Bern in August 1985, news reached us from Kathmandu that a major *jökulhlaup* had discharged from the Langmoche Glacier in the western Khumbu on 4 August. Daniel Vuichard and Markus Zimmermann had recently returned from completing the bedrock geology mapping and I was also in Bern. After consultation with all concerned, we telephoned Walther Manshard in Tokyo and immediately obtained a supplementary budget item for Daniel and Markus to return to Nepal and undertake a full field investigation of the glacier and the downstream impacts of the disaster. The implications of this will be related.

77   Khadga Basnet and Narendra Khanal, who had spent a year with me in Boulder as Nepali UNU fellows, joined Alton's team along with Sherpa locals, Khancha Lama and Pembra Sherpa. Because the methods to be employed were unfamiliar to Alton and I was not free to accompany the field party, I was relieved to persuade Colin Thorn to return to Nepal for a second tour and take my place.

78   Early in the observation season, Alton detected that several of the rain gauges contained liquid when no precipitation had occurred. Some careful detective work determined that small Sherpa boys had been mischievously contributing urine to the gauges.

79   Teiji Watanabe joined me in Colorado as a doctoral candidate immediately after Alton had completed, so they didn't actually meet until much later (2007) although they were working on closely parallel field problems. Much of Teiji's doctoral work took him to the Langtang Himal where he completed an impressive array of soil erosion and slope movement measurements for comparison with a parallel set in his home mountains of the Japanese Alps. He also took a lead in the study of Imja Lake and its potential for catastrophic outbreak (*jökulhlaup*).

80   I had assumed that these were photographs taken by Fritz in 1956 when he was a member of the Swiss expedition to climb Everest and Lhotse and that he had taken them with a phototheodolite on loan from Erwin Schneider. Alton believes that they were taken by Erwin himself. He may be correct. However, it was Fritz's material and the freak of Konrad's rescue and placing them in my care that ensured they eventually emerged as

critical glaciological evidence.

81 Several small airstrips, with weather stations, had been set up during World War II so that, in the event of engine trouble, U.S. aircraft, flying from Bengal across "The Hump" (the Hengduan Mountains) to supply the Chinese Nationalist (Kuomintang), had a chance of survival. Of course, the U.S. military was promptly withdrawn in 1947 although, fortunately for us, the remnants of one of the airstrips was well situated to be made into our 1985 base camp.

82 It was normal practice for official visitors at that time to be accompanied by security staff, although the extreme youthfulness of our two soldiers left us speculating.

83 Today the etymologically incorrect name that is "officially"used is Tiger Leaping Gorge. Photographs of it appear in numerous websites as a world class tourist attraction. The name of the entire gorge in the Naxi language is *A cai ggoq*, while its lower reaches carried the Chinese name *xia hu tiao*. Thank goodness, we managed to visit in 1985 and 1992 while it was still barely known outside the Lijiang region, although it had been traversed by Joseph Rock in the 1920s and 1930s.

84 Regretfully, this part of the research plan was a partial failure. The young, inexperienced Chinese assistants failed to maintain observations during periods of heavy rain.

85 Age-dating the trees, for instance, provided proof that mature forests had developed on areas that appeared on Rock's photographs as treeless. The great age of many subalpine forest stands (often in excess of 500 years) demonstrated that large areas had remained untouched by human tree-felling for centuries.

86 On the 1985 visit, the lake bed was carpeted in green, providing an important source of grazing for the village livestock.

87 A highly practical gesture was an invitation to indulge ourselves in the town baths. When Alton and I presented ourselves the following morning, we were marched through the streets in procession with the governor, cheering crowds lining the sidewalks. It was overwhelming. Then we found that, to ensure privacy, the entire and very large public bath setup had been scrupulously scraped clean and everyone moved out until we had finished.

88 After years of unsuccessful search (long ago I had lost my original Sunday School prize) I obtained a second-hand copy of *Storms on the Labrador* through the Internet. It arrived in August 2004. As I opened, it I recognized with enthusiasm the end-cover drawings. This was followed by dismay as I realized that the book did not contain a single reference to Sir Wilfred Grenfell, to the Grenfell Mission, nor to the Moravians of the Labrador northern coast! My only explanation is that the prize was awarded to coincide with my Sunday School teacher's series of lessons to augment the church's fundraising efforts for the mission. Thus the title of the book and the name of Sir Wilfred Grenfell had merged in my young mind.

89 Ivan Head was president of the Canadian International Development Research Centre and one of the closest advisors to Prime Minister Pierre Elliot Trudeau. Maurice was also very close to the Canadian prime minister.

90 This was due to the constitutional relationship between the Canadian federal and provincial governments, education being the preserve of the provincial governments.

91 The mounted photographs were later delivered for permanent display at UNU headquarters in Tokyo. I learned later that "they had proved so attractive that most of them, unfortunately, have been taken home by various members of our Board of Directors."

92 The proceedings were published as *Study Week on a Modern Approach to the Protection of the Environment*, Pontificiae Academiae Scientiarum Scripta Varia, 75, (Ed. G.B. Marini-Bettolo), 1989.

93 One especially memorable aside during the meeting resulted from an impatient interjection I made on the third morning. We had discussed major issues facing the world, including food shortage and poverty. I raised the need to add to the agenda the problems of uncontrollable population growth and birth control. This produced a shocked reaction prompting the chairman to insist on an immediate adjournment and early coffee break.

Each morning, the walls either side of the conference table were lined with robed observers. One of them immediately accosted me and took me into a small anteroom. I apologized for the disturbance I had caused and said that I had probably forfeited my personal meeting with the Pope. Instead, I was thanked for bringing up a vital and necessary major issue. The priest (who was Canadian and science adviser to the Pope)) went on to assure me that I would receive a special welcome "for courage," although His Holiness probably would not mention the topic of birth control. And so my high hopes were fulfilled; the photograph of His Holiness shaking my hand has a prominent place in my study.

94   Peter Stone had served for many years in Geneva as founding editor of the UN publication *Development Forum* and had been director of information for the Brundtland Commission.

95   *Mountain Agenda* was used as a name of convenience. There was no constitution, no officers, only use of the Institute of Geography, Bern University, as its address. In this way, we were signalling that it was open to all interested individuals and organizations, either formal or informal, with its initial purpose of bringing the mountain issue to widespread attention during the Rio Earth Summit.

96   The 300 copies of the book, endorsed by David Brower, Prince Sadruddin Aga Khan, Maurice Strong, and Lord Hunt of Llanfair Waterdine, together with an equal number of the *Appeal*, reached Rio a few days before the conference, courtesy of SwissAir.

97   The actual title of Chapter 13 was "Managing Fragile Ecosystems: Sustainable Mountain Development."

98   Dr Tewolde Berhan Gebre Egziabher had been a close mountain colleague since our 1980s meetings in Addis Ababa, Ethiopia, when we had helped establish the African Mountains Association, of which he served as president. In 2000, he was awarded the Right Livelihood Award, often referred to as the "Alternate Nobel Prize." In 2006, he received the UN top environmental prize: Champion of the Earth.

99   This shocked my UNESCO friends who had to remain behind the glass barriers as observers – the situation of all UN agency staff.

100   Another book was justified, in part, because *Status of the World's Mountains* had been produced under great pressure with no time for peer review. In practice, *Mountains of the World: A Global Priority* ,1997, edited by Messerli, B., and Ives, J.D., turned into an even more challenging task, although there was a fair element of peer review.

101   Sean's Chinese name is Xia Shanquan.

102   On our first visit in 1993 we received a hearty welcome and quickly learned that a special village feast had been prepared. The traditional initiation of such an event was the presentation of a very large pig. Its throat was carefully cut and its blood gently squeezed out into a large iron bowl while the poor animal screamed. Eventually the animal died, but not before the bowl was nearly full. I was then presented the bowl. I took a deep draft of the warm blood, passing it on to the elder on my left. It slowly passed through the circle of elders, all of whom followed suit and imbibed. We then began the feast with the entire village.

103   The following year, a group of Naxi school children made a visit to the Queen Charlotte Islands, homeland of the Haida.

104   On my final visit to Wenhua in November 1995, I hiked in with an education inspection team from Beijing. I learned from their leader that she intended to award scholarships to the two most outstanding and needy students (by this she meant money to ensure that they had shoes, clothes, and supplies such as pencils and notebooks, essential for them to remain at school). On asking the monetary value of a scholarship, I was astounded to learn that these "life-saving" awards amounted to the interest that US$200 would earn annually. I asked if it would be appropriate for me to provide funds for two more, to which she laughed in agreement. So I mentioned my intention to the head teacher and left the official delegation simply to make four awards instead of two, but on condition that my contribution was anonymous. Little chance! – my very modest contribution was reported, not only in the Lijiang newspaper, but in the main paper in Kunming. However, I enjoyed the ceremony and

thought no more about it. On my return journey home, some days were required in Kunming. Fuquan told me that we had been invited to dinner by a very important figure – General He Guocai, the senior military officer for Yunnan Province. It was a marvellous affair. He presented his card with a flourish and asked Fuquan to translate for me the Naxi script on the reverse side. It read: "I do not like to eat dogs." He was Naxi and that phrase was omitted from the Mandarin text on the front side. During the dinner, when I asked him why he had honoured me with such a magnificent banquet, he explained that he had read the Kunming newspaper. As a very poor boy, he used to walk barefoot to the Wenhai school. He then thanked me, promised me a military escort for the following year if I wanted to trek the then very dangerous far western gorges and urged me to use his assistance and purchase land on the shore of the Wenhai Lake and build my summer cottage there. He assured me that the way China was developing, it would become a very valuable property. I was not able to accept either offer, although his estimate of land and property values have proven themselves quite accurate. Yet the story remains one of those invaluable treasures of work with mountain people.

105 Sohrab and Rustum were the legendary characters of a Persian epic. Matthew Arnold, the English nineteenth-century poet, captured the agony and poignancy of a day-long single combat beside the River Oxus (Amu Darya) wherein the great warrior Rustum unwittingly slew his long-lost son.

106 For a variety of reasons I was not able to accept.

107 *Forests and Floods: Drowning in fiction or thriving on facts?* Forest Perspectives 2, Bangkok, Thailand, and Bogor Barat, Indonesia. FAO Regional Office for Asia and the Pacific and Centre for International Forestry (CIFOR, 2005).

108 I was invited to dinner in Kathmandu (February 2010) by Professor Hasnain when he discussed his Himalayan glaciological research and his denial – I have no reason to doubt his word.

109 The Mountain Institute (West Virginia, U.S.A.) has recently received extensive support from the USAID to develop a combination of social and scientific research on high mountain watershed problems in relation to climate change. This led to the expedition to Imja Lake, quoted above, and a continuation during the 2012 post-monsoon season. On the 2012 occasion an education and training workshop in association with the local people led to acquisition of additional scientific data from Imja Lake and highly promising Sherpa participation. Amongst the scientific results, sonar-based bathymetric investigation revealed that the volume of water contained within Imja Lake is twice that previously reported (65 M cubic metres) and the glacier front retreated much more rapidly over the preceding three months than previously expected. The "High Mountain Glacier Watershed Program", funded by USAID, is co-managed by The Mountain Institute and the University of Texas at Austin (Alton Byers, pers. comm., 25th November, 2012).

Index

Printed in the United States
by Baker & Taylor Publisher Services